JAGUAR WORLD

Jaguar 6 Cylinder Engine Overhaul (1948-1986)
(Including I.R.S. and S.U. Carburettors).

Published by
KELSEY PUBLISHING LTD

Printed in Singapore by Stamford Press PTE Ltd
on behalf of
Kelsey Publishing Ltd
Cudham Tithe Barn
Berry's Hill
Cudham
Kent TN16 3AG
Tel: 01959 541444
Fax: 01959 541400
E-mail: books@kelsey.co.uk
Website: www.kelsey-books.com

© 1995
1st reprint 1997
2nd reprint 2001
3rd reprint 2006
ISBN 1 873098 32 4

Acknowledgements

Our thanks go to Jim Patten for covering all the work undertaken in this book.

Our appreciation also to Gosnay's Engineering, Alan Slawson and Burlen Fuel Systems Ltd for carrying out the work described.

Contents

Engine
XK Engine-origins and definitions 8
Removal and dismantling. 14
The block 20
Cylinder head. 25
Timing gear and pistons 36
Assembly 40
Swapping cylinder heads and camshafts. 46
Tune-up 51

I.R.S.
Removal 54
Strip down 57
Dismantling components 60
Fitting bearings and adjustments 64
Stoppers and calipers 68
Differentials and discs 72
Hub carriers, driveshafts and refitting 75

SU carburettors
Strip down 80
Overhauling and reassembly 83

Your Jaguar deserves the best

JAGUAR WORLD
MAGAZINE

- buyer's guides
- archives
- accessories
- restoration
- maintenance
- news and events
- road tests

To subscribe, or for further details, go online to:
www.jaguar-world.com or call: 01959 541444

Introduction

I bought my first Jaguar, a 3.8 Mk2, at the age of 18 and I can still remember the unparalleled thrill of actually owning a 'Jag'. But this feeling was slightly tinged by an incredible fear of something breaking in what seemed to me then, a very complex engine, demanding at least a mechanic well-versed in race car technology. Of course it did fail otherwise there would be no tale to tell here, an exhaust valve had burnt and the cylinder head gasket was leaking. Oozing with enthusiasm and a rudimentary tool kit, I had a go. It took a number of months to sort out with more than one mistake on the way. The feeling of satisfaction when the engine, rebuilt with my own hands, actually started at the first push of the button cannot possibly be described here.

Our intention with the series of articles (fully compiled in this book) is to provide a step by step guide showing all the pitfalls and alternatives overlooked in the workshop manual. Whether you are a complete novice or are quite 'handy with a spanner', you will be able to tackle the XK engine with complete confidence. Just to make the story complete, we have included the SU carburettors and the independent rear suspension as well. Both are treated with the same thoroughness as the engine feature.

Of all the classic engines, the XK unit must rank as one of the finest in history. What other manufacturer can boast of such a diversity of usage? From luxury car to Le Mans-winning sports racer. From limousine to the Scorpion tank and its derivatives. Sports cars, funeral carriages, fire engines, ambulances, trucks and even record-breaking power boats and hydroplanes, all have their version of the XK unit. Built in the various guises of, 2.4, 2.8, 3.4, 3.8 and 4.2 litres in production as well as 3.0 and 4.5 litre in other racing applications, the range is truly remarkable, especially when you consider the extraordinary interchangeability of the parts. It shook the world when it was introduced at the 1948 Motor Show in the XK120 and many mourn its departure in 1992 after 44 years when the last DS420 limousine left the Jaguar works.

By following our guide you will be instrumental in ensuring that our beloved XK engine lives for ever.

Jim Patten.

Jaguar Quarterly, Winter 1990

The XK ENGINE –

Earliest application for the XK engine was in the XK 120; its blend of power, torque and silence was unmatched at the time. Note that this very early engine, in an aluminium-bodied car, lacks the front camshaft cover studs fitted soon after to control oil leaks.

By the late 1930s William Lyons was at a turning-point: he desperately needed more power if his ideas for a new range of luxury cars were to succeed. His chief engineer William Heynes (see *Jaguar Quarterly* Winter 89 and Spring 90) had squeezed 160bhp from the existing push-rod 3½-litre unit but only with considerable modification and, while performance was spectacular when fitted into the SS 100, it could not be sustained for more than a few minutes – the old unit had effectively come to the end of its development. However, it was decided to use its 160bhp as the base power target, with the additional requirement that there should still be plenty of scope left for further development.

Heynes, enthusiastic about twin-overhead camshafts and hemispherical heads ever since watching racing motorcycles at Brooklands, sketched out some basic ideas for his own twin ohc design during those famous firewatching evenings in war-time Coventry and it is some indication of Lyons' trust in his chief engineer that he eventually gave him the full go-ahead to design

Its overhaul and care

Part One – origins and definitions
By Jim Patten

and develop such an advanced power unit. Towards the end of the war, and assisted by Walter Hassan and Claude Baily, various experimental units were built including one with valve gear based on the cross push-rod BMW 328 engine. Eventually the classic twin ohc, hemi-head XK engine emerged, with, after a last-minute stroke increase to boost low-speed torque, an optimum bore and stroke of 83 x 106mm. This gave 3442cc in six-cylinder form and delivered the required 160bhp with a maximum torque of 195lb ft – quite enough to power the 'test bed' XK 120 sports car (in which the new

One of the series of experimental engines made prior to the finalisation of the XK engine's specification. This is the 1360cc XF, built to prove aspects of the ohc valve gear.

Two-litre, four-cylinder versions of the engine were made, Lyons envisaging that not enough demand would exist for the 3.4-litre alone. In fact, the company were overwhelmed with orders for the bigger engine and no four-cylinder cars were ever sold.

engine was announced on October 22 1948) to over 130mph

Serious consideration was also given to a four-cylinder unit of just under 2-litres and, indeed, much of the cylinder head development by gas-flow expert Harry Weslake was carried out on this engine. Jaguar even went as far as to catalogue this alternative as the XK 100 alongside the showstopping XK 120 in 1948, but the overwhelming – and totally unexpected – demand for the larger-engined car, especially from the US, ensured that the XK 100 would never reach production. It was not until 1955 that a smaller capacity engine arrived, when the 2.4 unit, a short stroke version of the 3.4, was fitted into Jaguar's new unitary construction compact saloon.

With the introduction of the XK engine Jaguar at a stroke entered the supercar league: no other

All XK engines were bench-run before installation, using coal gas. This picture was taken, probably early in 1950, at Jaguar's original factory in Swallow Road, Holbrooks, Coventry.

Model	Prefix	Engine Size Bore & Stroke	Cylinder Head Type	Valve Size In. Ex.	Camshaft Lift	Compression Ratio*	Block Type	Carburettor Type & Size	Flywheel	Clutch	Oil Pump	Timing Chain Tensioner	BHP
XK120	W WF Later	3442cc 83 x 106mm	A C option	1¾ 1 7/16	5/16″	8:1		2 x 1¾″ SU H6	132 teeth	9⅞ (s)	Gear Type	Spring Blade	160
XK120SE	W, WF Late '53	3442cc 83 x 106mm	A C option	1¾ 1 7/16	⅜″	8:1		2 x 1¾″ SU H6	132 teeth	9⅞ (s)	Gear Type	Spring Blade	190 210 C
MkVII	A, AB BD, D	3442cc 83 x 106mm	A	1¾ 1 7/16	5/16″	8:1		2 x 1¾″ SU H6	132 teeth	9⅞ (s)	Gear Type	Spring Blade	160
MkVIIM	D, DN NA	3442cc 83 x 106mm	A	1¾ 1 7/16	⅜″	8:1		2 x 1¾″ SU H6	132 teeth	9⅞ (s)	Gear Type	Spring Blade	190
XK140	G	3442cc 83 x 106mm	A	1¾ 1 7/16	5/16″	8:1		2 x 1¾″ SU H6	132 teeth	9⅞ (s)	From G1908 Rotor Type	Hydraulic From G4431	190
XK140SE	G	3442cc 83 x 106mm	A C option	1¾ 1 7/16 A 1⅝ C	⅜″	8:1		2 x 1¾″ SU H6	132 teeth	9⅞ (s)	Rotor	Hydraulic From G4431	190A 210 C
2.4 Saloon	BB	2483cc 83 x 76.5mm	A	1¾ 1 7/16	5/16″	8:1		2 x 24mm Solex Downdraught	104 teeth	9 (s)	Rotor	Hydraulic	112
2.4SE	BB, BC BE	2483cc 83 x 76.5mm	A	1¾ 1 7/16	5/16″	8:1		2 x 24mm Solex Downdraught	104 teeth	9 (s)	Rotor	Hydraulic	112
3.4 Saloon	KE KF	3442cc 83 x 106mm	B	1¾ 1⅝	⅜″	8:1		2 x 1¾″ SU HD6	104 teeth	9⅞ (s)	Rotor	Hydraulic	210
MkVIII	N NA	3442cc 83 x 106mm	B	1¾ 1⅝	⅜″	8:1		2 x 1¾″ SU HD6	132 teeth	9⅞ (s)	Rotor	Hydraulic	210
XK150 3.4	V	3442cc 83 x 106mm	B	1¾ 1⅝	⅜″	8:1		2 x 1¾″ SU HD6	132 teeth	9⅞ (s)	Rotor	Hydraulic	190
XK150 3.4SE	V	3442cc 83 x 106mm	B	1¾ 1⅝	⅜″	8:1		2 x 1¾″ SU HD6	132 teeth	9⅞ (s)	Rotor	Hydraulic	210
XK150 3.4S	VS	3442cc 83 x 106mm	Straight Port	1¾ 1⅝	⅜″	9:1		3 x 2″ SU HD8	Lightened 132 teeth	9⅞ (s)	Rotor	Hydraulic	250
XK150 3.8	VA	3781cc 87 x 106mm	B	1¾ 1⅝	⅜″	8:1		2 x 1¾″ SU HD6	132 teeth	9⅞ (s)	Rotor	Hydraulic	210
XK150 3.8S	VAS	3781cc 87 x 106mm	Straight Port	1¾ 1⅝	⅜″	9:1		3 x 2″ SU HD8	Lightened 132 teeth	9⅞ (s)	Rotor	Hydraulic	265
MkIX	NL	3781cc 87 x 106mm	B	1¾ 1⅝	⅜″	8:1		2 x 1¾″ SU HD6	132 teeth	9⅞ (s)	Rotor	Hydraulic	220
Mk2 2.4	BG, BH BJ	2483cc 83 x 76.5mm	B	1¾ 1⅝	⅜″	8:1		2 x 24mm Solex Downdraught	104 teeth	9 (s)	Rotor	Hydraulic	120
Mk2 3.4	KG, KH KJ	3442cc 83 x 106mm	B	1¾ 1⅝	⅜″	8:1		2 x 1¾″ SU HD6	104 teeth	9⅞ (s), 9½ (d) after KS8237	Rotor	Hydraulic	210
Mk2 3.8	LA, LB LC, LE	3781cc 87 x 106mm	B	1¾ 1⅝	⅜″	8:1		2 x 1¾″ SU HD6	104 teeth	1959-64 9⅞ (s) -'67 9½ (d)	Rotor	Hydraulic	220
240	7J	2483cc 83 x 76.5mm	Straight Port	1¾ 1⅝	⅜″	8:1		2 x 1¾″ SU HS6	104 teeth	8½ (d)	Rotor	Hydraulic	133
340	7J	3442cc 83 x 106mm	Straight Port	1¾ 1⅝	⅜″	8:1		2 x 1¾″ SU HD6	104 teeth	9½ (d)	Rotor	Hydraulic	210
E-Type 3.8	R RA	3781cc 87 x 106mm	Straight Port	1¾ 1⅝	⅜″	9:1		3 x 2″ SU ND8	Lightened 104 teeth	9⅞ (s) 9½ (d) after RA5801	Rotor	Hydraulic	265
E-Type 4.2	7E	4235cc 92.7 x 106mm	Straight Port	1¾ 1⅝	⅜″	9:1	Staggered Bores	3 x 2″ SU HD8	Lightened 133 teeth	9½ (d)	Rotor	Hydraulic	265
E-Type 4.2 SII	7R	4235cc 92.7 x 106mm	Straight Port	1¾ 1⅝	⅜″	9:1	Staggered Bores	3 x 2″ SU HD8	Lightened 133 teeth	9½ (d)	Rotor	Hydraulic	265
MkX 3.8	2B	3781cc 87 x 106mm	Straight Port	1¾ 1⅝	⅜″	8:1		3 x 2″ SU HD8	104 teeth	9⅞ (s)	Rotor	Hydraulic	265
MkX 4.2	7D	4235cc 92.7 x 106mm	Straight Port	1¾ 1⅝	⅜″	8:1	Staggered Bores	3 x 2″ SU HD8	133 teeth	9½ (d)	Rotor	Hydraulic	265
420G	7D	4235cc 92.7 x 106mm	Straight Port	1¾ 1⅝	⅜″	8:1	Staggered Bores	3 x 2″ SU HD8	133 teeth	9½ (d)	Rotor	Hydraulic	265
S-Type 3.4	7B	3442cc 83 x 106mm	B	1¾ 1⅝	⅜″	8:1		2 x 1¾″ SU HD6	104 teeth	-'64 9⅞ (s) after 9½ (d)	Rotor	Hydraulic	210
S-Type 3.8	7B	3781cc 87 x 106mm	B	1¾ 1⅝	⅜″	8:1		2 x 1¾″ SU HD6	104 teeth	9½ (d)	Rotor	Hydraulic	220
420	7D	4235cc 92.7 x 106mm	Straight Port	1¾ 1⅝	⅜″	8:1	Staggered Bores	2 x 2″ SU HD8	133 teeth	9½ (d)	Rotor	Hydraulic	245
XJ6 4.2	7L	4235cc 92.7 x 106mm	Straight Port	1¾ 1⅝	⅜″	9:1	Stg. Bores L/Stud	2 x 2″ SU HD8	133 teeth	9½ (d)	Rotor	Hydraulic	245
XJ6 2.8	7G	2792cc 83 x 86mm	Straight Port	1¾ 1⅝	⅜″	9:1	Equidistant Bores	2 x 2″ SU HD8	104 teeth	9 (d)	Rotor	Hydraulic	180
XJ6 4.2 S2	8L	4235cc 92.7 x 106mm	Straight Port	1¾ 1⅝	⅜″	9:1	Stg. Bores L/Stud	2 x 2″ SU HD8	133 teeth	9½ (d)	Rotor	Hydraulic	245
XJ6 3.4 S2	7L 8A	3442cc 83 x 106mm	Straight Port	1¾ 1⅝	⅜″	8.8:1	Staggered Bores	2 x 1¾″ SU HS6	133 teeth	9½ (d)	Rotor	Hydraulic	161 DIN
XJ6 4.2 S3	8L	4235cc 92.7 x 106mm	Straight Port	1⅞ 1⅝	⅜″	8.7:1	Stg. Bores L/Stud	Lucas Elec. F/injection	133 teeth	9½ (d)	Rotor	Hydraulic	170 DIN
XJ6 3.4 S3	8A	3442cc 83 x 106mm	Straight Port	1¾ 1⅝	⅜″	8.8:1	Stg. Bores L/Stud	2 x 1¾″ SU HS6	133 teeth	9½ (d)	Rotor	Hydraulic	161 DIN

* Standard UK

© WORLD COPYRIGHT PJ PUBLISHING LTD.

The XK ENGINE

The triple carburettor 3.8-litre engine perhaps represents the aesthetic pinnacle of the XK power unit – especially when exposed fully to view in the 'E' type.

manufacturer could offer such a performance engine as a production unit, with a specification that formerly could be found only in expensive bespoke sports-racing cars. Yet while the XK engine was quiet, docile and flexible in luxury car applications, between 1951 and 1957 it powered five Le Mans winners – and afterwards went on to power such diverse vehicles as limousines

It must not be forgotten that a majority of XK engines went to the United States in both saloons and sports cars. This is a Series 1 4.2-litre, again in an 'E' type and representing a useful comparison with the right-hand-drive installation.

and light tanks – not to mention record-breaking hydroplanes and race-winning power boats. But above all it powered the vast majority of production Jaguars built until the Series 3 XJ6 left production, thus seeing the company through their renaissance under John Egan and into privatisation.

Throughout its life the XK unit was developed, enlarged, reduced and fettled into its final form as the fuel injected 4.2 of the Series 3 XJ6. It still refuses to die, giving stalwart service today in the Daimler limousine.

It says much for Heynes' basic design that many parts can be interchanged between the first and the last units, although the final expression of the engine is a very different animal from that first fitted in the XK120. Interchangeability is a vast subject worthy of a book of its own but, during the coming months, this is an aspect which we will be constantly touching on. It is, for instance, possible to use a large valve, Series 3 straight-port cylinder head on an XK120, provided you do the right things!

Our chart gives a detailed explanation of engine types, specifications and model applications but, to help set the scene, we'll run through some of the major points of interest here. In its original form the engine appeared with the small valve, low camshaft lift 'A' type cylinder head, spring blade timing chain tensioner and a gear type oil pump. The advent of the 'C' type competion car inspired a 'performance' production cylinder head known as the 'C' type, available as a further option on Special Equipment XK 120s and XK 140s (by the time the latter arrived it carried a 'C' cast into the plug well). This head could be ordered by part number and is occasionally seen fitted to such as the Mk VII as a special order.

Slightly modified, the production 'C' head was to become the 'B' type, a standard fitment on many Jaguar models. Just to confuse things there was also the 'D' type head, essentially a big-valve version of the production 'C' head, used in standard form for the 'production' D' type sports racing cars (and thus for the short-lived XKSS derivative). Later there was what is termed the 'wide angle' racing head (some twin-plug examples of which exist). At one time Bill Heynes had slated the wide-angle for production use and, while this never happened, the production 'D' type head could be, and occasionally was, ordered separately under a part number.

The XK150 and Mk IX introduced the 3.8, created from an overbored, linered 3.4-type block, while the XK150S in 3.4 and 3.8 guise had the straight-port head fitted. As

S.E. IDENTIFICATION

'S' prefix to XK120 chassis number denotes special equipment model.

'A' prefix to XK140 chassis number denotes special equipment model with standard cylinder head.

'S' prefix to XK140 chassis number denotes special equipment model with C-Type cylinder head (also denoted by suffix 'S' on engine).

-7, -8, -9 after engine number denotes compression ratio.

NOTE: 2.4 SE did not involve any mechanical change. Cylinder head colours – all will be revealed in the cylinder head section.

Key to clutch: (s) sprung type (d) diaphragm type.

The XK ENGINE

mentioned earlier the compact saloon received the first 2.4 engine with 'A' type head. The 3.8 'S' engine (not to be confused with the 'S' type saloon which was never fitted with a straight-port head!) was carried forward in slightly modified form to the original Mk X and 'E' type of 1961.

In the never-ending search for more torque, the bores were enlarged to give 4.2-litres, accommodated by slightly staggering the cylinders in the block; all 4.2 engines had the straight-port cylinder head and (naturally) a new crankshaft carrying a different damper. The 240 and 340 saloons both received the straight-port cylinder head but, oddly enough, the 'S' type saloon in parallel production kept the 'B' type head, although by then its replacement the 420 had arrived. If you think Jaguar's model policy in the late sixties was messy, you're not the only one!

The XJ6 of 1968 adopted what was virtually the 420's two-carburettor 4.2-litre engine and, in common with late 420G saloon and 'E' type, had a modified cylinder block with head studs extending down to the crankshaft main bearing area. The XJ6 equivalent to the erstwhile 2.4-litre was the 2.8; this, too, had the straight port head but the bores were equidistant. However, when the next variant, a 'reborn' 3.4, joined the XJ Series 2 range the bores were offset in the same way as the 4.2 – though the bore/stroke dimensions remained the same as the original XK 3.4-litre's (useful for early 3.4 rebuilds as its lighter, higher compression pistons will fit).

All XJ6 blocks and heads were

Less known is the use of the XK engine in commercial and military vehicles; Dennis used it briefly for an ambulance and it is still employed today in the Scorpion light tank.

The XK engine lived to power the most successful Jaguar of all time, the XJ6. By this time the engine had been internally and externally updated but was essentially the same design as had emerged in 1948.

slightly extended to the rear to accommodate an additional cooling passage, while the camshaft profiles were altered for quieter running. The Series 3 saw the XK's first and only inlet valve size change when these were enlarged to what had been the production 'D' type diameter of 1⅞in. For the first time on a road Jaguar, too, fuel injection replaced carburettors, mainly for emission control reasons.

This over-view represents a brief outline of the XK engine's evolution – its numerous variations will be examined in greater detail in the issues ahead as we come to talk about the individual types of engine and the way they are overhauled.

It is likely that Daimler limousine production will cease at the end of 1991 and with it will go the very last of the XK units from car production – truly a legend laid to rest. During its immensely successful life span, somewhere in the region of 650,000 production engines were built and it is not over-stating the case to say that the XK engine was Jaguar's greatest asset and mainstay for over 35 years. Without Lyons' foresight and willingness to gamble on an engine with such an advanced specification, and the competance of the Heynes/Hasan/Baily team which designed it, maybe Jaguar would never have survived as a major force in the luxury-car market.

NEXT ISSUE

XK engine dissection!
Removing and dismantling the power unit.

If only ...
HE HAD BOUGHT HIS PARTS FROM HENLYS

- We stock only genuine Jaguar parts.
- Our link to Jaguar's centralised stock of parts means that we can get most parts within 24 hours
- Genuine parts will undoubtedly save you money in the long run
- Daily deliveries in London
- Special discounts for members of Jaguar Enthusiasts Club and Jaguar Drivers Club
Mail Order Welcome

Henlys (London) Ltd., The Hyde, Edgware Road, London NW9 6NE

Telephone: Simon or Les on Parts Hotline
Tel: 0181-200 8363
Fax: 0181-200 9939

PARTS OPENING HOURS
Mon-Fri 8.30 - 5.30
Sat 9 - 1

HENLYS JAGUAR
Est. 1928

JAGUAR GENUINE PARTS — Daimler OFFICIAL DEALER

JUST JAGS

JUST JAGS HEADLAMPS JUST JAGS EXHAUSTS
JUST JAGS ECU'S JUST JAGS DIFFERENTIALS JUST JAGS NEW USED AND RECON PARTS
JUST JAGS SPRINGS JUST JAGS BRAKE PADS JUST JAGS BRAKE DISCS
JUST JAGS WATERPUMPS JUST JAGS STEERING RACKS

LATE JAGUAR NEW RECONDITIONED AND USED SPARES AVAILABLE FOR:-
JAGUAR XJS, XJ40, XJ6, S3
PRIVATE, TRADE, OVERSEAS ENQUIRIES WELCOME

Large stocks of Used Spares
Tel/Fax: **01922 37779**
Mobile: **0831 506722**
Mail Order Service Available
Late damaged/undamaged Jaguars wanted

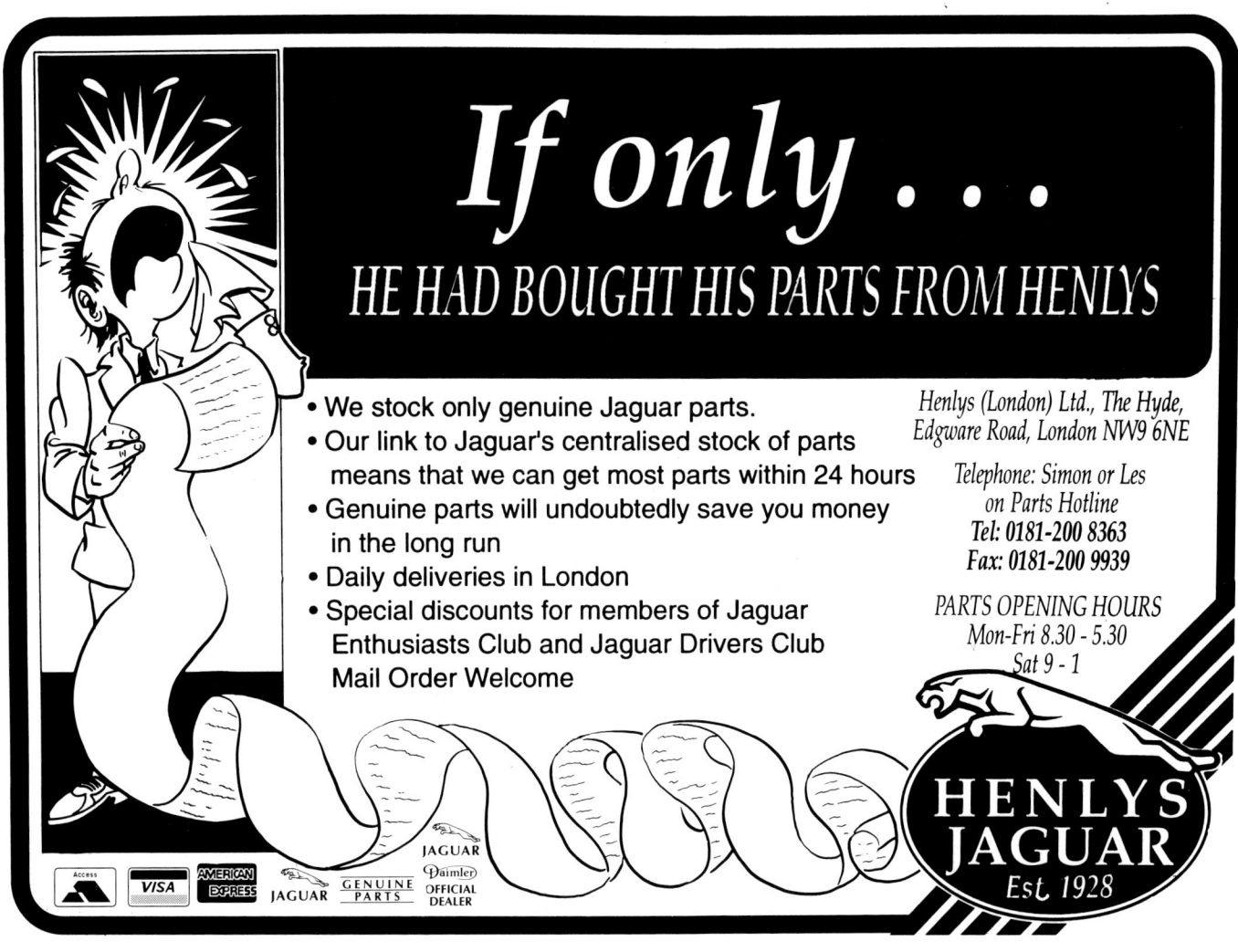

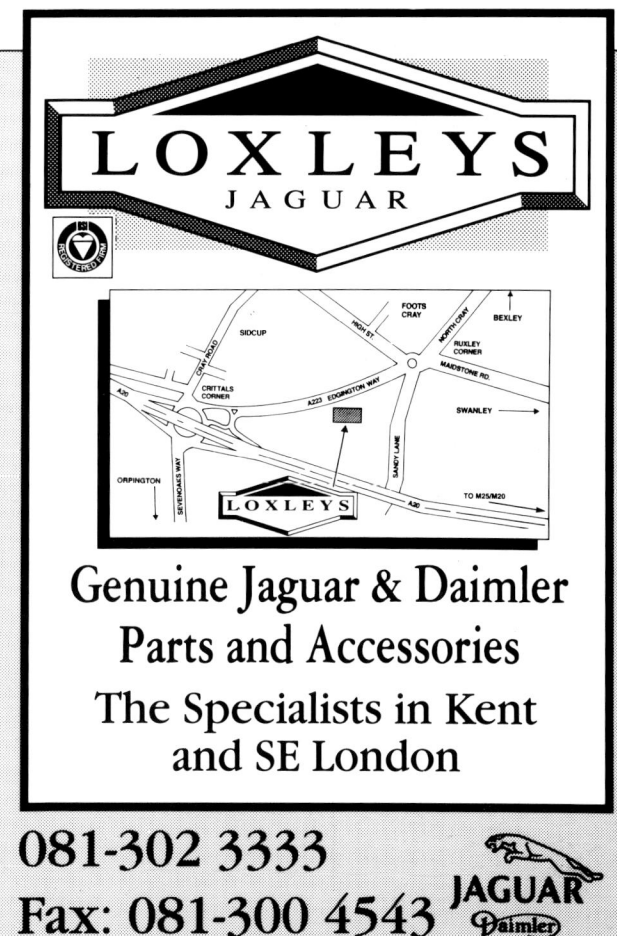

LOXLEYS JAGUAR

Genuine Jaguar & Daimler Parts and Accessories

The Specialists in Kent and SE London

081-302 3333
Fax: 081-300 4543

JAGUAR Daimler

Edgington Way • Sidcup • Kent DA14 5BN

Jaguar Quarterly, Spring 1991

The XK ENGINE –

Part Two – removal and dismantling

By Jim Patten

We now move on to the removal and initial dismantling of a typical XK engine, as told by the picture sequence accompanying. Throughout this guide we will be referring either to Jaguar's official workshop manual or that published by Autobooks – though other publishers, such as Haynes, also usefully supplement Jaguar's own literature.

Preparation

This can't be over emphasised and it pays to plan the operation thoroughly. You'll find it much easier, for instance, if the engine and bay are steam cleaned (or degreased by hand) first. Then it sounds obvious, but make sure that you have a number of jars or boxes ready for nuts and bolts, plus felt-tipped pen and labels to help identification on reassembly.

Read and digest the procedures involved a few evenings in advance as it does help to understand what you are going to do beforehand. You will need physical help as well – Jaguar components are very heavy and are not to be managed alone. The engine hoist must be man enough for the job, rated to carry at least 8cwt; there have been some very serious injuries due to using inadequate equipment.

TOOLS REQUIRED

UNF socket and spanner set.
3/8in Whitworth socket (for top timing chain adjuster).
Top timing chain adjuster.
7/32, 3/16 & 1/2in Allen keys.
Circlip pliers.
Feeler guages.
Torque wrench.
Camshaft timing plate.
1 5/16in AF socket (for the crankshaft nut).
1/2in crows foot spanner (for timing chain slipper pads).

---

*Engine removal by the book, which does allow you to leave carbs and manifolds in **situ** if required – but note the steep angle involved and how frames and bodywork need protecting. Photo courtesy **Practical Classics**.*

The writer in a happy moment removing the engine from beneath OCH 887E during its restoration. In this case two chain hoists were used to lift the car, using nylon strops on each hub.

Its overhaul and care

DISMANTLING THE ENGINE

Remove all chrome dome nuts securing the camshaft covers and breather housing and lift the covers away.

Release the tension on the top timing chain by loosening the nut on the eccentric idler and, by using the special tool (see workshop manual), turn in a clockwise direction. Then snip and remove the locking wire between the two bolts holding the sprockets to the camshafts.

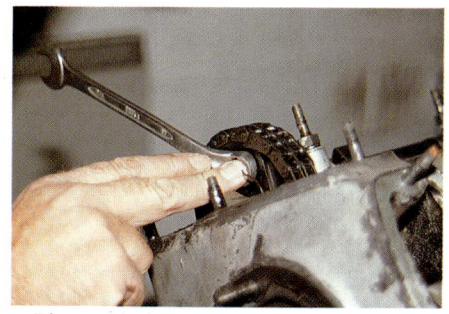

The camshaft bolts can then be removed one at a time on each camshaft, rotating the engine for access. Ease the sprockets off the camshaft and slide up the support brackets. Don't turn the engine after this as damage to the valves could occur.

Slacken the 14 chrome dome cylinder head nuts before finally removing them; to be absolutely correct, undo in reverse of tightening order.

Don't forget the six nuts around front of cylinder head.

The book will tell you to lift off the cylinder head; don't (necessarily) believe it…

…this one required a bottle jack between the head flange and the block flange to get any movement…

…and then various blocks of wood used as levers…..

Engine Removal 'E' type

Strictly speaking, it isn't necessary to remove the bonnet if the engine is removed from below, it can just be swung forward and supported but, realistically, this is a last resort. Bear in mind that the bonnet forms virtually a third of the car and is extremely heavy – reserve a safe place to store it.

Most manuals recommend that the engine and box are lifted out from above; there is nothing wrong with this but we tend to favour lowering the combined units to the floor on to a waiting trolley. This makes it much easier to manoeuvre and minimises the possibility of damage. It's quite easy to fabricate a trolley from some Dexion angle, wood and castors. Refer to the manual for the conventional method – here, we describe the alternative.

With the bonnet out of the way, place the car on a substantial pair of axle stands forward of the engine, rear wheels firmly chocked. Drain all fluids from the engine and then, referring to your manual, remove the radiator and all engine ancillaries – dynamo/alternator, distributor, water pump, carburettors and inlet manifold, exhaust manifolds and down pipes, oil filter. Make sure that every wire, linkage, piping, fuel line and all other connections between engine and car are removed.

You should now see the basic engine stripped in its frame. From inside the car remove the seats, gearlever knob, radio console and armrest (where fitted). The gearbox top cover, gearlever and reversing light switch wiring can now be removed. From underneath the car disconnect the speedometer cable and the propshaft from the gearbox flange. It will not be necessary to disturb the hydraulic pipe from the clutch slave cylinder – just remove the two bolts securing the body to the bellhousing and lift it away.

There's a reaction tie-plate between the rear of the front torsion bars that passes between the engine and gearbox, and this must be removed. The workshop manual describes a rather messy procedure using silver-steel bars and levers. Follow this if you wish but we favour relieving tension on the torsion bar by releasing the front suspension top ball joint. The shock absorber must be in place as this limits the suspension travel; otherwise the ball-joint will 'neck'.

It will now be possible to undo the four nuts and bolts securing the torsion bar to the tie plate. *Don't disturb the torsion bar bracket on its*

Jaguar Quarterly, Spring 1991

The XK ENGINE

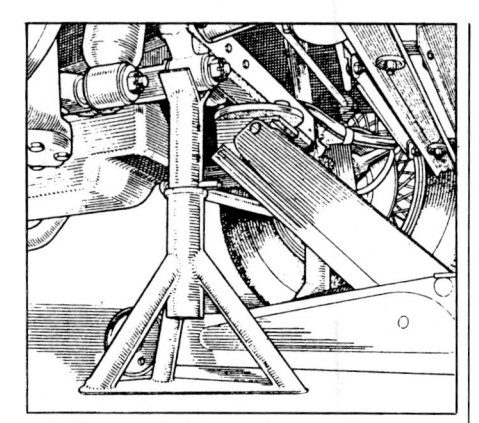

Make sure that the car is secured on substantial axle stands. Always use a piece of hardwood between the jack and the frame.

splines as the torsion bar will then need resetting. After the bolt passing through the side of the tie plate and underframe channel is removed the tie plate can be tapped off.

Bear in mind that, if the above method is adopted, the car will be immovable because, without the engine weight, it will be impossible to reconnect the top ball joint.

The engine will now be left on its two front mounts, the top stabiliser and the two gearbox 'cotton reel' mounts on the support plate. Most later Jaguars have lifting eyes secured between the cylinder head bolts; if yours hasn't there really are bucket fulls in scrapyards. Alternatively, a chain could always be passed around the engine girth.

Place an engine sling between the lifting eyes and position the crane to take the weight in a balanced manner. Our purpose-built engine/gearbox transporter can now be placed to receive its load. Remove first the engine stabiliser followed by the gearbox mount. To be safe, place a trolley jack under the gearbox.

Now undo the front engine mounts and the engine brackets – essential, otherwise the engine will not clear the frames. Check finally that everything is disconnected and, when you are absolutely sure, lower the jack and remove it, then lower the engine on to the trolley with the crane. Your assistant should be looking for snags or that overlooked wire. With the engine firmly on the trolley, the front of the car can be raised until there is sufficient clearance to wheel the whole lot away. An 'E' type without its engine is not too heavy and can be easily lifted – use the engine crane with a nylon sling around the front frames.

...and finally the engine crane before the head parted with the block! During these operations check that the gasket isn't fouling as this can add to your troubles. Remove the studs from the block – two nuts locked together and turned from the bottom one will work if you don't have a stud extractor. Note that looking from the front, on the second row the righthand stud is dowel type.

Remove clutch by undoing bolts securing cover plate to flywheel, knock back the lock tabs on the flywheel bolts and remove. The engine will need to be locked for this operation – a stout screwdriver wedged in the ring gear usually works.

Just a few words about removing engines from other models. The XJ6 range is possibly the easiest and there are no hidden snags. With the bonnet off and radiator out, the unit can come out through the front, with or without the gearbox. On the Mk 2/S-type range again we favour extraction from beneath. Although this entails the complete removal of the front suspension crossmember it's not as hideous an operation as you may

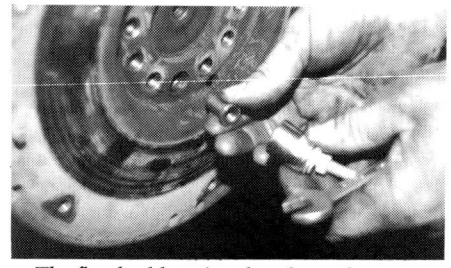

The flywheel locating dowels can be removed by using a long nut and bolt; place a spacer (with a sufficiently large diameter to enable the dowel to pass inside) between dowel and nut, screw the bolt into the dowel, then wind the nut down on to the spacer. The dowels will be pulled out by this method.

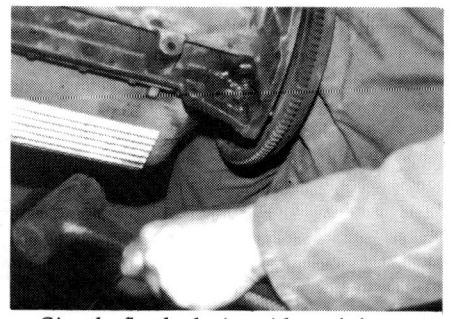

Give the flywheel a jar with a soft face mallet and ease away.

BUILDERS OF THE FINEST JAGUAR ENGINES IN THE WORLD

German Historic GT

Bridgestone Challenge

3.8 series 1

World's Fastest E-Type

Post Historic Touring Cars

V12 Series III

Suppliers of numerous engines for concours winning cars including:

Several class winners at the Benson & Hedges Classics

Car of the show at JDC National Day for two years running

Car of the show at JEC Northern National Day

Best XK at the International XK Day for two years running

ROB BEERE ENGINEERING

Unit 9, Herald Business Park, Golden Acres Lane, Binley, Coventry CV3 2SY
Tel: 01203 650521   Mobile: 0860 675001   Fax: 01203 650539

H·A·FOX
GUILDFORD

JAGUAR Daimler APPROVED USED CARS

JAGUAR Daimler OFFICIAL DEALER

JAGUAR
SPECIALIST PARTS SUPPLIER

ENTHUSIASTS — MAIL ORDER A SPECIALITY

OWNERS   FAX: 0483 505832

• DIRECT COMPUTER LINKS TO JAGUAR FOR SLOWER MOVING PARTS

• PAYMENT BY CREDIT ACCOUNT (Application Form by Request)
Proforma Invoice, Charge Card (Access, Barclaycard or American Express)

We carry comprehensive stocks of genuine factory-approved and fully warranted parts.
Discount for members of the JE Club.

**1 Ladymead
Guildford
GU1 1DT**

01483 65207

Ring the Specialist Parts Hotline

01483 570196

Remove the front pulley and undo the crankshaft damper bolt. The damper will probably need a jolt with a soft faced mallet to free it.

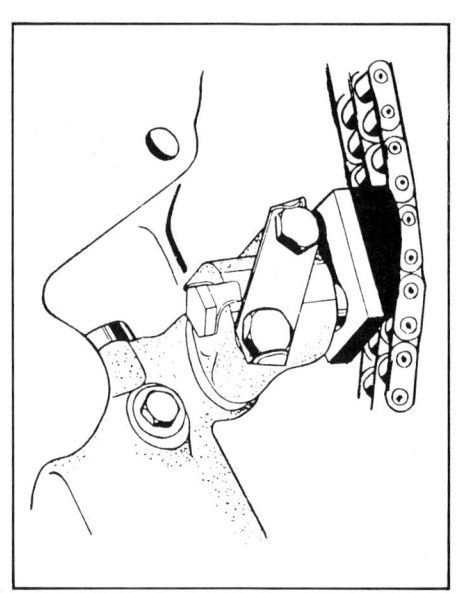

Turn the engine over for access to the sump in such a way that the timing gear is undisturbed. Undo the 26 bolts and four nuts securing the sump; note that there is a shorter bolt to the right-hand front corner of the sump.

The hydraulic tensioner on this engine was so worn that it dropped into the sump.

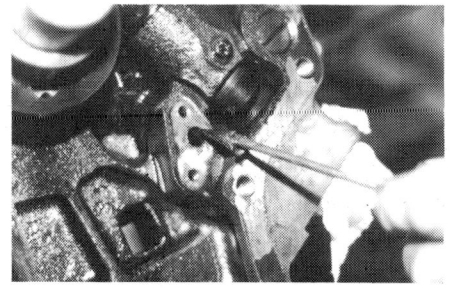

The conical filter behind the tensioner can now be removed and soaked in cleaning agent.

Remove timing chain cover to expose timing chain assembly. Undo the hexagonal plug in end of hydraulic tensioner, insert a .125in AF Allen key into the restraint cylinder and turn clockwise to free slipper head from the chain (unless the slipper head has fallen into the sump, in which case this operation is unecessary!). The two bolts can then be undone and the tensioner removed.

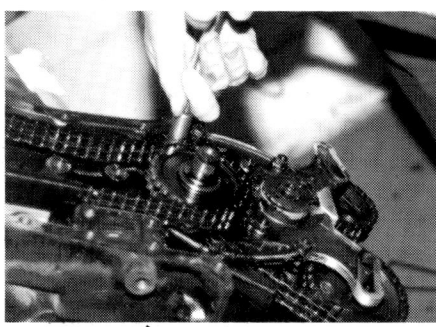

The four bolts holding the timing chain front mounting bracket and the two screwdriver slot setscrews securing the rear mounting bracket can now be removed and the whole assembly lifted off.

The bolt heads are irritatingly concealed on the vibration damper and a well-cranked spanner or one of special construction will be needed to remove it.

think – it means disconnecting only four mounts, the anti-roll bar, steering column and the brake pipe union. By placing a piece of plastic sheet between the brake reservoir and filler cap, effectively blanking out the breather hole, there should be no fluid loss and only the front brakes will need bleeding.

With the engine in a convenient place, thoroughly degrease using Gunk, Jizer or similar. Petrol works very well but is just too hazardous in a confined area. Remove the starter motor and undo all the gearbox bellhousing bolts and lift the gearbox away. Don't underestimate the weight of the gearbox – it's a very heavy item and will need support during this operation. The engine is now ready to be dismantled.

Jaguar Quarterly, Spring 1991

The XK ENGINE

Remove the oil delivery and pick-up pipes before undoing the four bolts securing oil pump to front main bearing cap.

Put the coupling piece in a safe place until needed.

Release the nut at bottom of oil pump drive shaft. Be very careful here as the distributor drive gear is on this shaft and meshes with the crankshaft gear; there's a chance of damage. An upturned 3/8 drive socket with an allen key in the socket end can be used to exert opposing pressure on the drive shaft, thus avoiding an unnecessary risk.

Tap the drive shaft free of the block and lift away the helical gear.

The pistons are now ready for removal. Release the big-end nuts (early examples have split pins). Note that caps and rods are numbered and are fitted one way only. Keep them together after removal.

Push the piston out through the block using a non-metallic bar. Keep all pistons in order, although they are numbered on the connecting rods.

The crankshaft can now be lifted away. That completes the initial dismantling and I think you'll agree that it wasn't too difficult. The cylinder head has been left intact as this will be dealt with in a seperate episode.

Use Allen keys to remove the bolts securing the cover assembly to the rear of the block.

Undo main bearing bolts and remove main bearing caps. Remember that these are handed and should be kept in order. Numbers are stamped on caps and adjacent on the block.

Our thanks to: Geoff Maycock, Autocats, Rawreth Industrial Estate, Rawreth Lane, Rayleigh, Essex SS6 9RL. Tel: 0268 782306.

NEXT ISSUE
How to identify component wear, and what goes on at the machine shop.

Jaguar Quarterly, Summer 1991

The XK ENGINE
The machine shop

**Part 3
Jim Patten
explores a secret
world**

Ladies and Gentleman, please step forward and bear witness as the mysteries and subtle secrets of the Engine Machine Shop are revealed!

In the entrance of our Engine Machine Shop lies two piles of engines. To the left are the dirty, oil-encrusted exhausted ones waitin' to start their journey of recovery through stages of rehabilitation by methods shrouded in mystery. The finished ones, cleaned, painted and revi'alised, are stacked to the right of the shop waitin' to rejoin the world of the livin'.

What happens in the darkness, what is that mechanical cacophony we hear? We 'as been allowed into the secret domain and can now reveal to you, ladies and gentlemen, that knowledge… Roll up, roll up!

Well, for all the average person knows about the machine shop it might just as well be a 19th century mystery show. So we called in to see Adrian Wilkes, director of Gosnay's Engineering of Romford, Essex to get an insight into the machine shop's secrets. Adrian has followed his father into the family business and has some 25 years of experience in engine machine work.

Gosnay's are game for pretty well anything and we saw examples of early Aston Martin, Bentley and Bugatti as well as more mundane Morris 8 engines, plus an occasional early aircraft unit thrown in for good measure. But they've rebuilt literally hundreds of Jaguar engines over the years and there isn't much about machining the XK power unit that they don't know.

The block is examined first, initially for cracks; these can occur across the bore centres or can frost-crack along the main sides. An engineer's straight-edge is placed along the top face and feeler gauges inserted to check for distortion. Any more than 0.003in usually spells trouble and really means resurfacing and line boring the main bearings. Don't forget to bolt on the timing chain cover when skimming.

HALF PRICE BOOKS®

Half Price Books
1835 Forms Drive
Carrollton, TX 75006
OFS OrderID   27206399

Thank you for your order, Krm #JB!

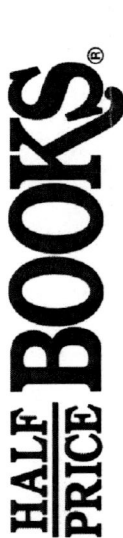

Thank you for shopping with Half Price Books! Please contact Support@hpb.com. if you have any questions, comments or concerns about your order (111-3574958-4486656)

Visit our stores to sell your books, music, movies games for cash.

SKU	ISBN/UPC	Title & Author/Artist	Shelf ID	Qty	OrderSKU
S373881599	9781873098325	Jaguar 6 Cylinder Engine Overhaul: 1948-19... Motorbooks International	33--12--7	1	

ORDER# 111-3574958-4486656
AmazonMarketplaceUS

SHIPPED STANDARD TO:
Krm #JB
839 MOASIS DR STE 3C
LITTLE CHUTE WI 54140-1229
qnr6wd6s02ly6d9@marketplace.amazon.com

Jaguar Quarterly, Summer 1991

Gosnay's use this line honer to ensure correct alignment of the main journals.

They are minutely lapped for true…

….and then checked for tolerance.

Should the block require line boring, then a small amount is taken off the bottom of the main bearing caps and the whole line bored on this enormous boring bar.

The bores can now be checked, first for any deep scores (surface scratches are lost in the rebore process). Then internal dimensions are checked with an internal micrometer. The measurement is taken fore and aft and then across the bore both at the top and bottom. If wear exceeds 0.006in a rebore will be necessary.

Core plugs should be removed and any build-up of sediment flushed. This becomes almost solid over the years and a good machine shop will soak the block in an acid mix or use a good degreasing agent – but all too many poor ones don't.

Here a block is about to be bored oversize.

The block is set on the bottom face and trued to the boring bar for precise machining. A measurement is taken from the top of the bore to ensure central location.

Excessively worn bores will require new liners. This enables the block to be brought back to standard size. XK engines prior to the 3.8 weren't linered, of course, but can be if the wear exceeds the rebore limit.

Today's oils make the bedding-in process more difficult, making it essential to prepare the bores for running by honing them. This process takes off any high spots or troughs left behind after the boring process. Gosnay's recommend a running-in oil after reconditioning.

The bore on the right is shown after honing compared to the untouched example on the left. If it's found that the bores do not exceed the wear limit and just new piston rings are to be fitted, then honing is still essential to remove the 'glaze' build-up on the bore.

It is worth checking across the centre main bearing as this should give an indication of any likely excess end-float.

Crankshafts take a terrible pounding; this one shows the obvious signs of burning where a bearing has 'picked up'.

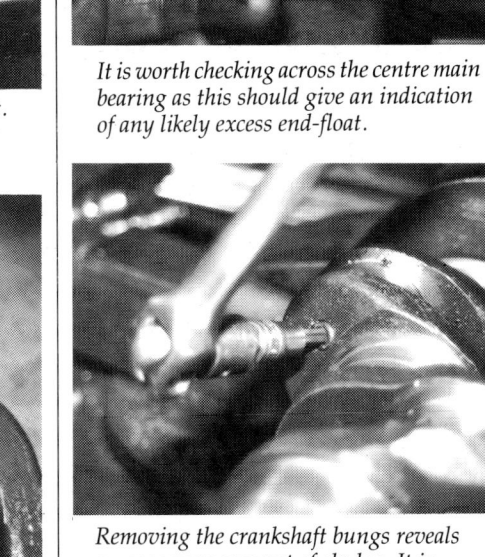

Removing the crankshaft bungs reveals an enormous amount of sludge. It is essential that this is removed as it will severely reduce oil flow. Because it is out of sight, many machine shops will not bother with this task. The bungs can be difficult to shift and sometimes need drilling out.

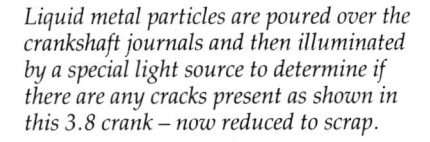

Liquid metal particles are poured over the crankshaft journals and then illuminated by a special light source to determine if there are any cracks present as shown in this 3.8 crank – now reduced to scrap.

All journals are now measured in two planes for wear. Any more than 0.003in wear requires a regrind of the crankshaft.

Thoroughbred Car Services

Independent Jaguar Specialists for over 20 years
Servicing – Repairs – Restoration – Tuning

All cylinder head repairs carried out – corroded waterways reclaimed – valve guides, tappet guides, valve seats etc. renewed.

Unleaded conversions & performance modifications, machining carried out on latest SERDI equipment.

NEW & USED PARTS FOR MOST MODELS

Full or partial restoration on all Jaguar models.

Steering, suspension, axle, gearbox overhauls.

Engine overhauls from standard to full race spec. Competition preparation.

Winners of best prepared car award CSCC/BARC Historic Touring Car Championship 1993.

31 Rutherford Close, Eastwood Industrial Estate, Leigh-on-Sea, Essex SS9 5LQ
Tel: 01702 520797 Service 01702 420844 Parts/Fax

GOSNAY'S ENGINEERING COMPANY LTD
ESTABLISHED 1935
MAWNEY ROAD TRAFFIC LIGHTS · EASTERN AVENUE · ROMFORD · ESSEX
Telephone: ROMFORD 01708 740668 · 743331 · 74820 Fax: 01708 733266   RM7 7NS

PRECISION ENGINEERS AND ENGINE RECONDITIONERS

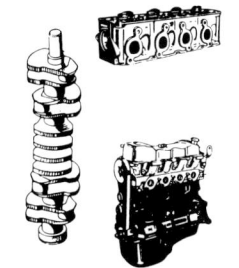

CYLINDER REBORING • CYLINDER RESLEEVING • IN-CHASSIS ON-SITE REBORING AND RESLEEVING • CRANKSHAFT REGRINDING • CRACK DETECTION • PRESSURE TESTING • WHITE METALLING MAIN BEARINGS AND CON RODS • SURFACE GRINDING • OHC CAMSHAFT HOUSINGS RECLAIMED • MACHINING OF CYLINDER HEAD FIRE GROOVES • VALVE AND VALVE SEAT REFACING • VALVE SEAT INSERTING • FITTING AND MACHINING OF CAM BUCKETS • LINE BORING • CON ROD RECONDITIONING

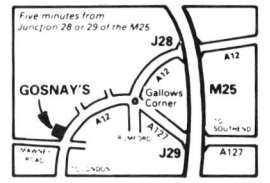

PISTONS • BEARINGS • GASKETS • VALVES/VALVE GUIDES • CYLINDER LINERS • RING GEARS • OIL PUMPS • CAMSHAFTS • CAMFOLLOWERS • TIMING CHAINS

EXCHANGE:- ENGINE UNITS • SHORT ASSEMBLIES • CRANKSHAFTS • CON RODS • CYLINDER HEADS • CAM SHAFTS • BRAKE DRUMS AND DISCS •

OUR STOCK OF ENGINE COMPONENTS IS SECOND TO NONE.
IF YOU REQUIRE PARTS FOR VEHICLES ANCIENT OR MODERN, PETROL OR DIESEL – GIVE US A CALL   24 HOUR DELIVERY SERVICE

CARRILLO

Finest quality steel con rods stocked for JAGUAR 3.8/4.2
Specials made to order
Engine test facilities up to 800BHP
Dynamometer Sales & Servicing, Dynamic Balancing

EDS ENGINE AND DYNAMOMETER SERVICES Ltd
Arisdale House, Arisdale Avenue, South Ockendon, Essex. RM15 5AS England
Tel: 01708 857108 Fax: 01708 855917

Ken Shergold

Specialising in Jaguars from XK's to XJ's. Small business with 27 years experience from giving advice to full restorations.

Three miles from Junction 25 (M25).

27 years of service to our clients at home and abroad.

KEN SHERGOLD
9 Fieldings Road, Cheshunt,
Hertfordshire EN8 9TY
Tel: 01992 624721 Fax: 01992 633499

Jaguar Quarterly, Summer 1991

The crankshaft is placed in an offset grinding lathe and set up to run on the journal centre lines so that...

...there can be no variance when each journal is ground.

Metal is removed from the crankshaft in areas dictated by the machine's scope to achieve balance.

With all components machined it's well worth having them balanced. Here the crankshaft is built up with the damper (make sure this is in top condition and if there are any doubts then replace – the crankshaft can break if not correctly damped), pulleys, flywheel and clutch assembly.

Pistons are weighed and machined to match, while the connecting rods are measured for length before they, too, are weighed.

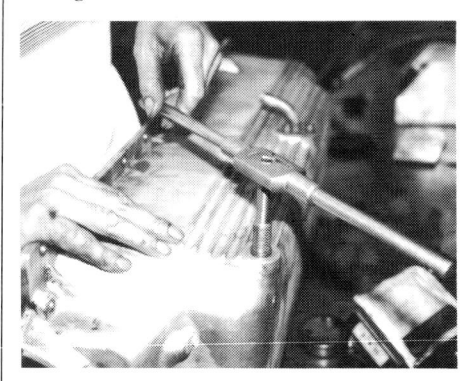

The thread on this sump drain plug had stripped, requiring a Heli-Coil to be fitted.

Your components should now be ready for collection and hopefully this insight into the machine shop should dispel any myths or doubts you may have had.

Our thanks go to:

Gosnay's Engineering Co Ltd, Mawney Road Junction, Eastern Avenue, Romford, Essex RM7 7NS (Tel: 0708 740668/748320).

Peter Trent of Thoroughbred Car Services, 31 Rutherford Close, Progress Road, Southend-on-Sea, Essex (Tel: 0702 520797), for the crankshaft illustrations.

EDS of South Ockendon, Essex (Tel: 0708 857108), for the balancing work.

NEXT ISSUE

Cylinder head overhaul

Jaguar Quarterly, Autumn 1991

The XK ENGINE

Part Four – Jim Patten begins the first of two instalments on the XK engine's crowning glory

Hemispherical combustion chambers in a light alloy head, topped by twin overhead camshafts and fuelled by multi carburettors. In 1948 this could only mean Maserati, Ferrari or any other exotic temperamental sports racer. But this was not a traffic jam splutterer, this was a Jaguar and the Coventry firm's intention was to apply this specification to a luxury saloon.

In a world of small capacities and sidevalves, the twin overhead camshaft XK Jaguar engine must have appeared advanced indeed and its very appearance sent shivers down the back of all but the most competent of mechanics who feared over-complication.

Today, even sporty hatchbacks boast twin ohc engines and the engineering fright has largely subsided. Most have at least grown used to higher levels of sophistication and are now more able to deal with an absence of push rods. As multi-valve engines become ever more common we may even eventually consider the XK engine as quaint.

That the XK engine is a most durable unit has been proven but if there is any weak link then the cylinder head must be it. The combination of an aluminium head with a cast iron block can produce dire corrosion problems, certainly when anti-freeze without its inhibiting properties is omitted. This, coupled with extreme temperature variances, will lead to

Slacken off camshaft bearing caps evenly....

....remove (note that caps are numbered to match the head location), and lift the camshaft away. Although not strictly necessary, it's worth tying a label to each camshaft to identify inlet and exhaust.

25

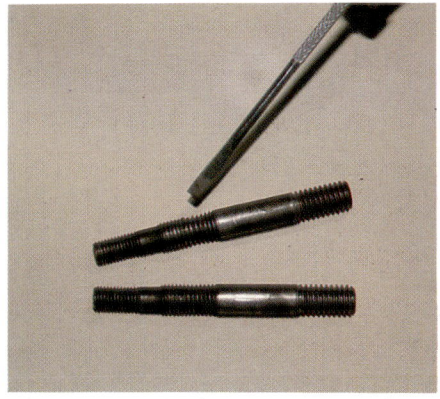

Remove all studs. At this stage, for what it costs, it is well worth replacing them. This example shows a very dubious thread. If re-using the original studs, roll them along a piece of card – this will show if they are bent in any way.

The tappet buckets and valve shims can now be removed. The easiest way is simply to use a magnet or the sucker on a valve grinding stick.

A conventional valve spring compressor can be used but it must have a deep end to fit in the valve spring recess.

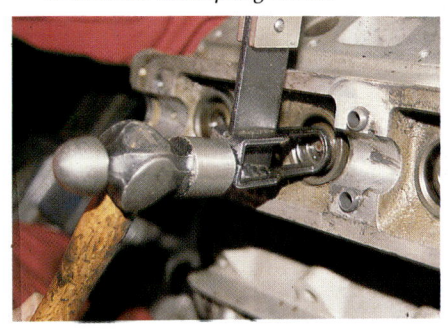

With the tool in place, give the end a light tap to free the spring from the valve stem.

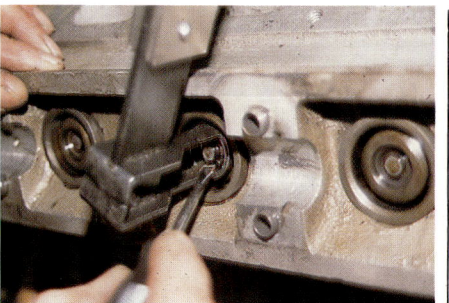

When the spring is fully compressed, the securing collets can be removed; a magnetic screwdriver is just the job.

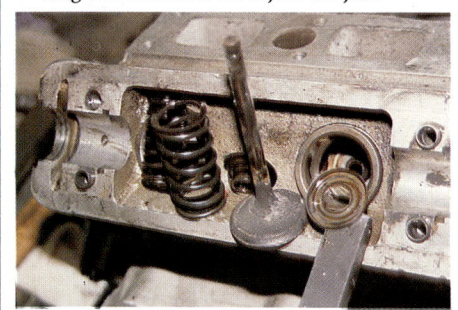

Valve, springs, collar, collets and spring seat can all now be removed. Continue this operation for all 12 valves.

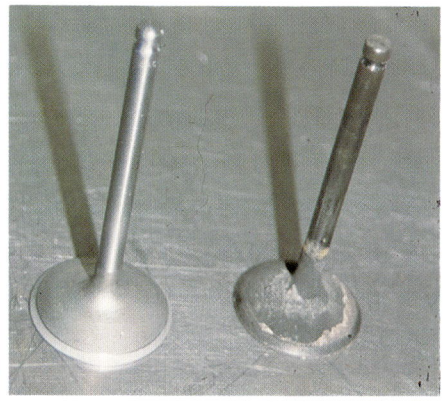

It is a false economy to attempt to re-use any of the internal parts. The used valve on the right will suffer from stem wear and diminished seat facing; springs will have sagged. Collets, collar and spring seat can be re-used.

Remove any core plugs by drilling through the centre and applying leverage. Thoroughly clean out behind.

Thoroughly clean the stripped head, making sure that every bit of carbon is removed. DO NOT use anything that will scratch the combustion chamber – avoid knives and screwdrivers. A final polish with wet and dry paper is perfectly permissible. At this stage the plug well can be painted the correct colour (see chart in next chapter).

Ensure that all stud holes are clear and, if possible, run a tap through all threads; check for any burrs.

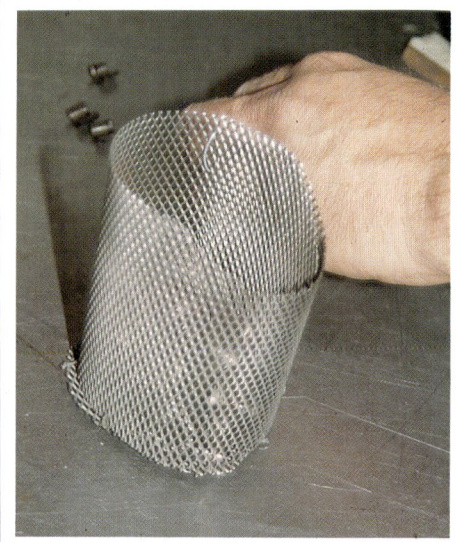

Collets, collars and spring seats can be cleaned by placing in a mesh basket and dipping in a tin of degreaser.

VALVE CLEARANCES

	Inlet	Exhaust
Up to engine nos.: 'E' type 7R8687/7R8854, XJ6 2.8 7G5794, XJ6 4.2 7L8343, 420G 7D60698.	0.004	0.006
After above engine nos.	0.012	0.014

If changing to later camshafts, note that they can only be fitted in pairs.

More power for your Jaguar?
For road or race, Prowess sets the standard

At Prowess Racing we develop and build ultimate Jaguar engines. Whether tuning a standard 3.8 E-type engine to give an effortless road-going 265bhp (the figure Jaguar always claimed for the unit, but failed by around 80bhp to realise), or creating what we believe to be the most powerful and reliable racing engines available, our painstaking methods have proved themselves time and time again. That is why during 1991/2 the Le Mans-winning Jaguar D-type XKD606 won 11 out of the 22 races it entered without needing any engine attention. It is also why at the Coys Festival at Silverstone in 1994 Barry Williams was able "to drive away from Sytner (Ferrari GTO) by over 3 seconds a lap with ease" to win comfortably in Nigel Corner's lightweight E-type. If you would like to discuss your Jaguar's performance potential call Hugo Spowers: we think you'll discover a car you never knew you had.

Prowess Racing (Restoration) Ltd, Windlesham, Surrey. GU20 6BX
Telephone: 01276 451428 Facsimile: 01276 451840
Winners of the Montagu Trophy, Pebble Beach Concours. 1993

F. B. Components
35-41 Edgeway Road, Marston, Oxford OX3 0HF

The complete range of **JAGUAR PARTS** for all models, including XK, MkI, MkII, S Type, 420, Mk10, E Type, XJ6, XJ12 & XJS.

* We have the biggest **FREE CATALOGUES** of available parts – send a large stamped addressed envelope stating model or call and see our large stock.
* We send parts all over the world to owners, enthusiasts' clubs, garages and dealers by post, air freight, sea freight and carrier.

Tel: 01865 724646   Fax: 01865 250065

Series One Engines JAGUAR SPECIALISTS

Full or part engine rebuilds, crank grinding, head & block surfacing.
All components supplied.
Service and repair, refinishing and restoration.
Over 20 years experience.

Series One Engines, Mill Farm, Whalebone Lane North, Chadwell Heath, Romford, Essex RM6 5QX.
Tel: 0181 597 3079 Fax: 01375 361370

The XK ENGINE

Machining is covered in the next issue, so assume at this stage that the valve seats and guides have been done. Smear some coarse valve grinding paste on the valve face, affix the valve grinding stick and with a regular motion roll the stick between the hands to 'lap' the valve into its seat. Continue this, adding paste when necessary, until there is a uniform grey hue around the valve face and seat. Clean off all traces of paste and start again using a fine grade. The seat and face will now show a slightly different shade of grey and will appear smoother. It is actually possible to gauge the operation by ear. At the start of grinding, the motion sounds coarse – then gradually eases.

The end result should be a continuous smooth grey band around both the valve face and seat. Make sure that the valves are clearly identified so that they are re-fitted to the correct seat.

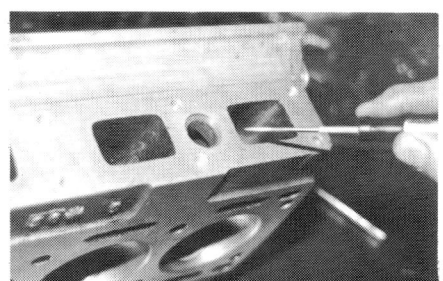

When all the valves have been ground, meticulously clean every part of the cylinder head, removing every particle of paste, much of which will have found its way inside the ports.

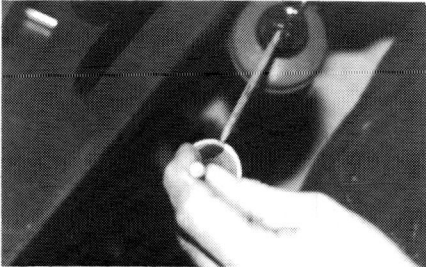

Apply a small amount of oil to the stem and assemble with the springs, spring seat and collar.

The springs can now be compressed and retaining collets fitted with the help of our magnetic screwdriver.

VALVE ADUSTING PADS

	inches	mm
A	0.085	2.16
B	0.086	2.18
C	0.087	2.21
D	0.088	2.23
E	0.089	2.26
F	0.090	2.29
G	0.091	2.31
H	0.092	2.34
I	0.093	2.36
J	0.094	2.39
K	0.095	2.41
L	0.096	2.44
M	0.097	2.46
N	0.098	2.49
O	0.099	2.51
P	0.100	2.54
Q	0.101	2.56
R	0.102	2.59
S	0.103	2.62
T	0.104	2.64
U	0.105	2.67
V	0.106	2.69
W	0.107	2.72
X	0.108	2.74
Y	0.109	2.77
Z	0.110	2.79

Adjustment shims are lettered with 'A' being the smallest. If possible, fit this size to all valves as it makes the adjustment calculation easier, but at £1.50 each shim it is expensive. Failing this, use the smallest size available and make a note of the valve number and size of shim fitted. Do not put oil on the shim as this could give a false reading.

Fit the tappet bucket with a light application of oil on the outside. Use only genuine Jaguar parts; the reproduction item on the left would not fit.

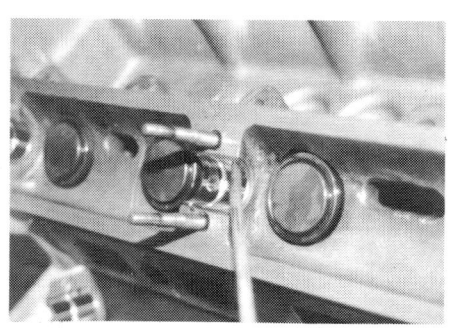

Fit new studs and camshaft bearings and apply a liberal coat of oil.

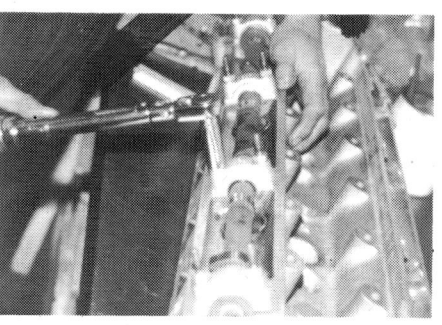

Re-fit the camshaft ensuring that the camshaft caps are fitted according to the corresponding numbers on the cap and head. Tighten evenly to a torque of 15psi. Rotate the camshaft several times to settle. FIT ONE CAMSHAFT ONLY – on all but the early small-valve engines the two together will result in valve collision when turning.

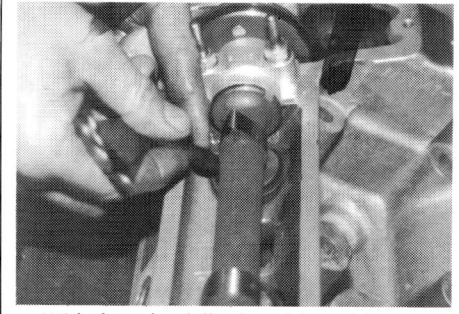

With the valve fully closed (cam lobe pointing up), measure the clearance and note....

gasket failure and head distortion. Fortunately, from an engineering standpoint (if not an economic one), just about every item of damage can be rectified.

For the purposes of this instalment we will be covering only the stripping and re-assembly of the head. As the machining is quite complex, this, and a full inspection, will be covered in the next issue.

Bring your bodywork to life

- Body & panel work
- Full or part restorations
- Insurance approved accident and chassis repairs
- Glasurit two-pack spraying to original colour specification
- Low bake ovens
- Rewiring & trimming - complete or partial
- Body modifications and conversions
- Servicing and repairs
- Mechanical overhauls
- Complete workshop facilities
- Inspections & valuations

Unique 3 year warranty on all work

The Ultimate Classic Motor Company

SOUTHERN CLASSICS

Expert repair and restoration of classic and quality cars

Hanworth Lane, Chertsey, Surrey KT16 9LA
Tel: (01932) 567671 Fax: (01932) 564482

NORTHAMPTON AUTORADS

ESTABLISHED 1967

THE AUTOMOBILE RADIATOR & FUEL TANK SPECIALISTS

A Traditional Company with Traditional Values

We offer a full range of Jaguar Radiators on request.

We also offer a complete high quality, restoration service for all types of car and commercial Radiators, Fuel Tanks & Heaters restoring them with specification parts.

Additionally we manufacture Radiators, and Fuel Tanks in Aluminium, Copper and Brass for any vehicle

For more information on our Company, please refer to the article in the November/December 1994 issue of Jaguar World.

Catalogue and Price Lists Available. Trade Enquiries Very Welcome.

51-53 ROBERT STREET NORTHAMPTON NN1 2NQ
TEL/FAX 01604 35937 - 30191

Access   VISA

The XK ENGINE

Head Case

Part Five – Jim Patten investigates a new pastime

"A Knurling we will go my deary oh, My inserts are all frozen, And my seats they are too low."

If you think that this is part of a mislaid script for Kenneth Williams in *Round the Horn*, then think again. We're back in the machine shop and attempting to decipher its fascinating linguistic code.

A typical cylinder head showing extensive corrosion around the waterways.

DAVID MANNERS
for all your
XJ PARTS & PANELS

THE COMPANY

15 years ago David Manners was having difficulty in obtaining parts for his Daimler Dart SP250 and realised that this could be the basis of a successful business.

He appreciated the difficulties experienced by owners of other Jaguar & Daimler models and has gradually extended the range of parts to cover: XJ40-XJ6-XJ12-XJS-MKII-S-Type-420-420G-Daimler V8-SP250 and DS420.

The company is conveniently located in the heart of the West Midlands, only 50 yards from junction 2 of the M5. Customers can visit the on-site parts counter or take advantage of the fast and efficient mail order service.

SALES & SERVICE

One of the many advantages that David Manners has to offer is the enormous stock of parts and panels carried at all times.

Due to a continuing investment, the business now carries over 11,000 different lines and as a result of its manufacturing facility, this range is continually increasing - anything from parts books and workshop manuals to complete assemblies.

Whether you are a retail customer looking for one item or a wholesale company interested in purchasing several thousand parts, David Manners will provide the service you require.

Not only will you be able to discuss your requirements with knowledgeable and helpful staff, but you will also be able to use the unrivalled mail order service (next day delivery if ordered before 3.00pm).
We can often offer original parts or alternative ones at discounted prices.

Free Price Lists are available on request.

MANUFACTURING

Over the last 7 years David Manners has invested time and money in building an on-site manufacturing facility. This has allowed the company to develop and produce parts and panels that were obsolete or difficult to obtain. We are currently manufacturing over 150 items.

XJ
The very competitively priced parts and panels that we produce include XJ front cross-members, floors, toe boards, radius arms, inner and outer sills and the coupe outer sill.

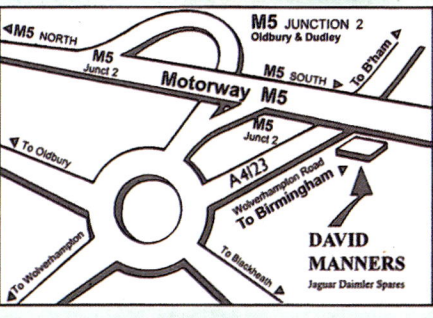

CUSTOMER CARE

Customer care is David Manners first priority. It is the factor above all others that has enabled the company to expand so successfully.

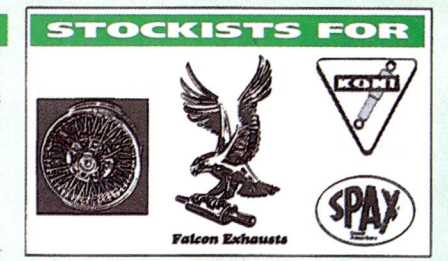

STOCKISTS FOR

KONI • Falcon Exhausts • SPAX

FREE PRICE LISTS FOR MOST MODELS

DAVID MANNERS
PARTS FOR JAGUAR & DAIMLER CARS
991 WOLVERHAMPTON ROAD, OLDBURY, WEST MIDLANDS, B69 4RJ. (50 Yards from M5 Junction 2)
TEL: 0121-544 4040 FAX: 0121-544 5558 / 0121-544 5888

Before welding can begin the waterways need to be cut back to a sound base.

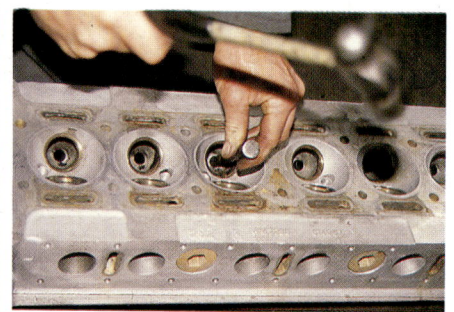

The cylinder head needs to be heated before the guides are driven out. This operation is repeated for all guides with some haste as the head will soon cool, making their removal impossible.

Here the internal dimensions of a pattern valve guide are checked. They are often incorrect, requiring honing to the right size after fitting. It pays to use original Jaguar parts.

If the valve guides are only slightly worn they can be reclaimed by a process known as knurling. A roller knurler is run down the guide and actually displaces metal in furrows to make up for wear. A diamond honer is then run down the guide to achieve the correct dimension. A maximum of 0.005in can be reclaimed in this way.

To remove the valve seats, the cylinder head must be accurately set for the machining process to be absolutely spot on.

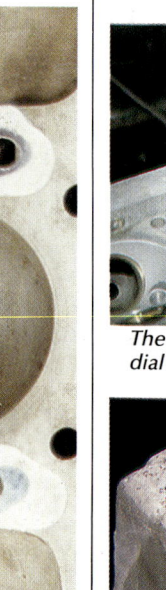

This illustration shows the repair after welding and an initial cut to size. If this was all that was required the head normally would be skimmed at this stage.

The cutting tool is first centralised with a dial gauge.....

.....and then the cutting tool moves in to remove the complete seat. Careful measurements are required to ensure that an extremely accurate fit is achieved.

This combustion chamber awaits removal of valve guides and seats.

The guide is replaced in the same way as the removal – the head is re-heated and the guide is then driven in.

The cylinder head is more susceptible to abuse than any other part of the engine. Using water alone without an inhibitor, usually present in anti-freeze, has dire corrosive effects on the aluminium used in the head. Overheating leading to severe distortion could easily reduce the head to scrap as, although the face could be skimmed to provide an

The XK ENGINE

The insert should be frozen using liquid nitrogen and then placed over a mandrel...

.....and driven into a machined space in the combustion chamber.

New valve guide and seat firmly in place.

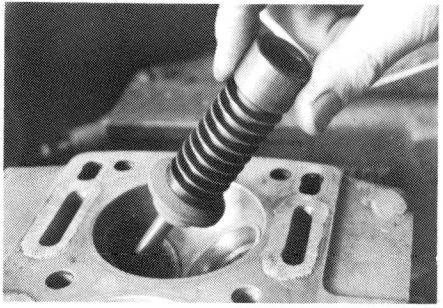

The seats now need to be cut ready for the valve grinding operation. The correct angle cutter is placed stem first in the guide for alignment.....

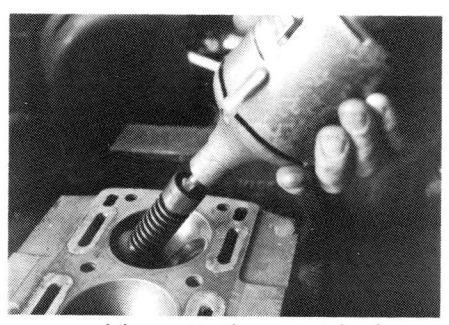

......and then cut to the correct depth; too much and the Jaguar shims will be robbed of clearance. Of course, the valve will later be lapped for final finishing.

The tappet guide is removed in much the same way as the valve seat. First the cutting tool is centralised and then the cutter moves in for an accurate removal.

More measuring as the tappet guide is checked aginst the machined space in the head.

Tappet guides frozen, head heated and the guides placed on a mandrel for rapid insertion before the temperatures normalise.

accurate mate to the cylinder block, the camshafts would no longer run true in the bearings, causing them to whip and wear the bearings. Today, however, the situation often can be remedied by specialist machining. We now look at the remedial work involved to bring an original head back to correct specification.

Remember that the work we are following here is that of a specialist machine shop and gives an indication of what can be done to retrieve a cylinder head. We're not suggesting that any of this specialist work can be done in the home workshop but feel the information to be useful for those wanting to

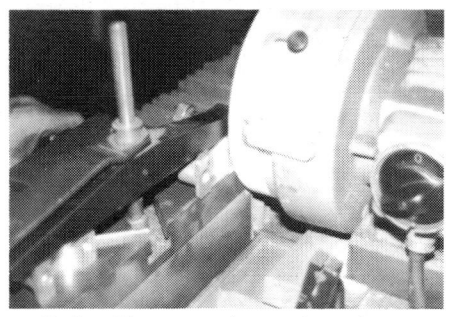

If a head has warped more than a few thou' then the camshaft bearing housing will need to be line bored to stop camshaft whip. First the caps are skimmed by a few thousands of an inch.

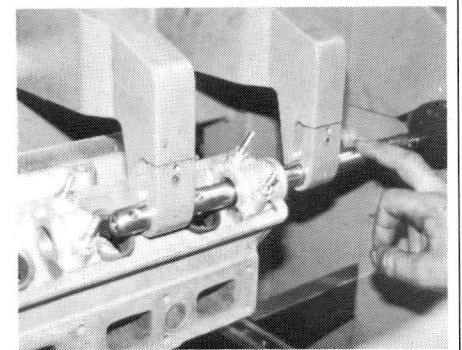

A datum point is taken along the bearing line and a cutter fed through supported by overhead pillars.

know exactly what happens through every step. You must really want to use a particular cylinder head to undertake all of this work because the expense can be horrendous.

Acknowledgements

The patience of Peter Trent of Thoroughbred Cars and Adrian Wilkes at Gosnay's is something to behold. They have put up with the idiotic antics of your scribe and have bitten their tongues before answering some absurd questions. To complete a job with a photographer crawling all over the work and then ask the operative to do it again takes a special kind of understanding. In my own garbled way, this is a thank you for two extremely helpful and co-operative people. Without their time and knowledge these features would not have been possible.

Peter Trent of Thoroughbred Car Services is at 31 Rutherford Close, Off Progress Road, Eastwood Industrial Estate, Leigh-on-Sea, Essex SS9 5LQ (Tel: 0702 520797).

Gosnay's Engineering Co Ltd are at Mawney Road Traffic Lights, Eastern Avenue, West Romford, Essex RM7 7NS (Tel: 0708 740668/743331).

NEXT ISSUE
Starting to put it all back together again.

Each bearing is cut in turn with supporting pillars holding the cutting tool. It has been impossible to carry out this operation in the past as the line boring required a support at each end of the tool, prohibited due to the height of the front of the cylinder head.

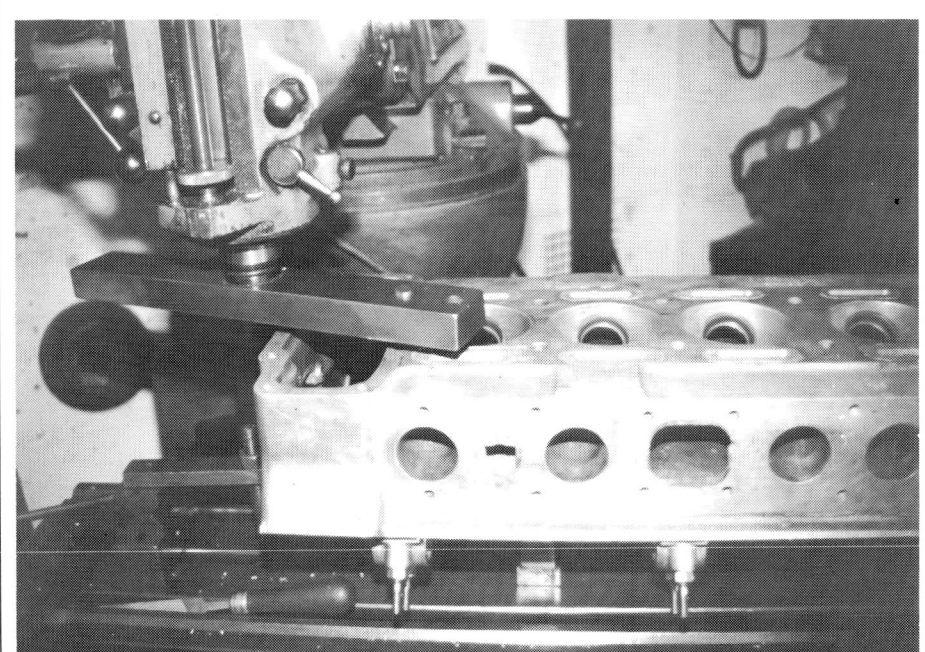

With all operations finally completed the head face can now be skimmed until true.

The resurfaced head face should now be a gas-tight fit on the engine block. Remember that, if the head has had major work through distortion, the cam cover mounting faces will also need resurfacing as these, too, will be warped and likely to leak.

Norman Motors Ltd

IF CHOOSING YOUR JAGUAR WAS EASY ...

... CHOOSING A DECENT JAGUAR SPECIALIST TO SUPPLY YOUR ENGINE & MECHANICAL PARTS CAN BE THE HARD BIT!

At Norman Motors we have over 20 years experience specialising only in Jaguar. Our small team of knowledgeable, polite staff can ensure that whatever the needs for your Jaguar – we can supply. Our huge stock range covers literally **everything** for Jaguar – both new and quality remanufactured – at competitive prices.

We are known throughout the world and our customers come back to us time and time again.

Norman Motors Ltd. 100 Mill Lane, London NW6. Tel: 0171-431 0940 Fax: 0171-794 5034
SHOP OPEN Mon-Fri 9am-6pm. Sat 9am-1pm.
MAIL ORDER SERVICE, EXPORTS AND SHIPPING ARRANGED TO ALL COUNTRIES

H.R. OWEN
TRADITION OF EXCELLENCE

For all of your Jaguar parts and accessories, call the genuine parts specialists.

COWDRAY AVENUE, COLCHESTER, ESSEX CO1 1DP
TEL: 01206 764764 FAX: 01206 761855

MALAYA GROUP PLC

The XK ENGINE

Part Six –
Final Preparations

By Jim Patten

We're approaching the final furlong – just a couple of jumps to go, the way ahead looks clear, and if the stamina lasts it could be a strong finish.

There's only some pre-assembling to deal with now before the engine as a whole can be put back together. It will be assumed that the timing gear has been dismantled as this is a very easy process.

Timing Gear

Timing chains are replaced as a matter of course. However, examine the gears themselves, looking for pointed teeth and wear between the teeth. Any detected and the gears should be replaced.

Look at the rubber dampers – there probably will be tram lines deep in the faces. Replace them all.

Remove the eccentric adjuster; in extreme cases the shaft can get locked with congealed oil. Gentle application of heat or immersion in boiling water will free it.

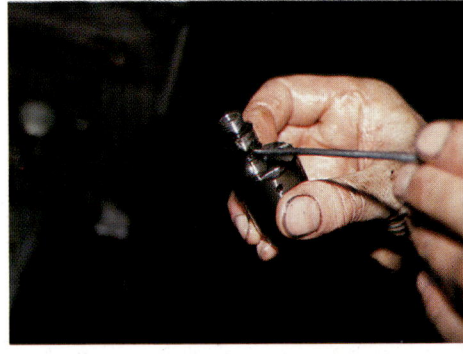

Check that the oil galleries are clear and remove the screw from the back for complete access.

Thoroughly clean in preparation for assembly, and make sure that all gears spin freely on their shafts.

Replace the eccentric shaft on to the rear bracket and place the intermediate sprocket in position; this is secured behind with a circlip.

The idler sprocket can now be fitted on to the eccentric shaft together with the top (longest) timing chain and camshaft sprockets. Follow the chain route as shown.

Then the bottom timing chain (shortest) is placed on the intermediate gear.

The front bracket can now be fitted; it is worth putting the dampers and spacers into position and temporarily securing them with nuts at the rear. An elastic band across the camshaft sprockets will keep them in place.

First place the spring and plunger pin in position by the idler sprocket and then secure the adjustment plate to the the shaft, remember that the nut is a Whitworth size. The assembly can now be placed aside until needed.

Oil Pump

Although new it is still essential to check the oil pump clearances. Using soft jaw covers in the vice, remove the pump rotor cover.

There should be not more than 0.010in between the outer rotor and pump body.....

..... and 0.006in between the inner and outer rotors.

End-float is checked by placing a straight edge across the face of the body and measurement taken by feelers between the rotors and straight edge. This should be 0.025in.

With the oil pump re-assembled, place the 'O' rings in the outlet and inlet holes.

The XK ENGINE

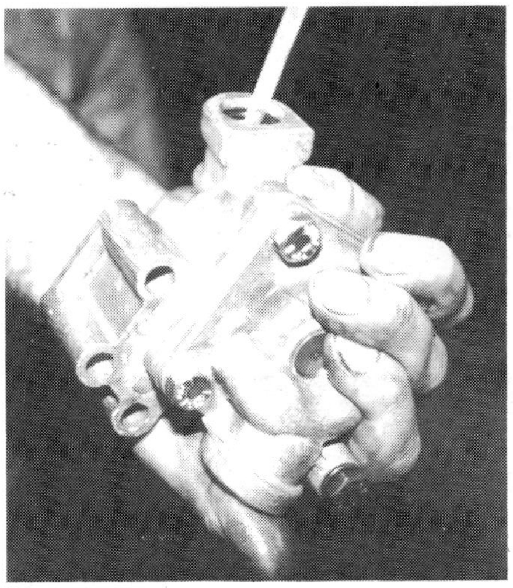

Prime the pump with engine oil.

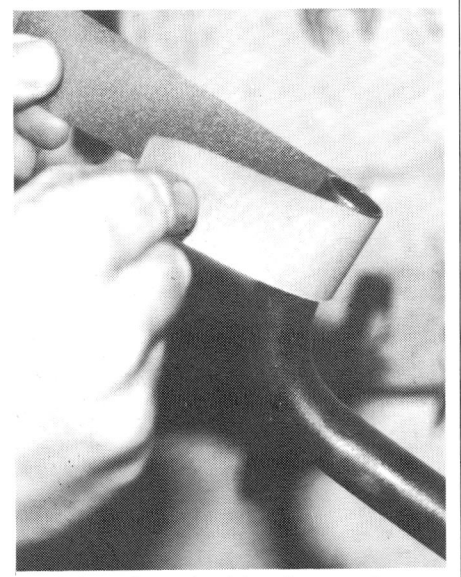

Polishing the ends of the oil pump pipes makes fitting a lot easier and makes for a better seal.

Pistons, sump, oil filter housing & crankshaft

To fit the piston to the connecting rod it will be necessary to heat the piston to free the gudgeon pin. Immersing in boiling water will do nicely. Make sure the piston is fitted the right way round. The size stamped on the piston crown will be at the front, while the connecting rod number should face the exhaust side of the engine. The small end bush should be coated with oil and then the gudgeon pin pushed through.

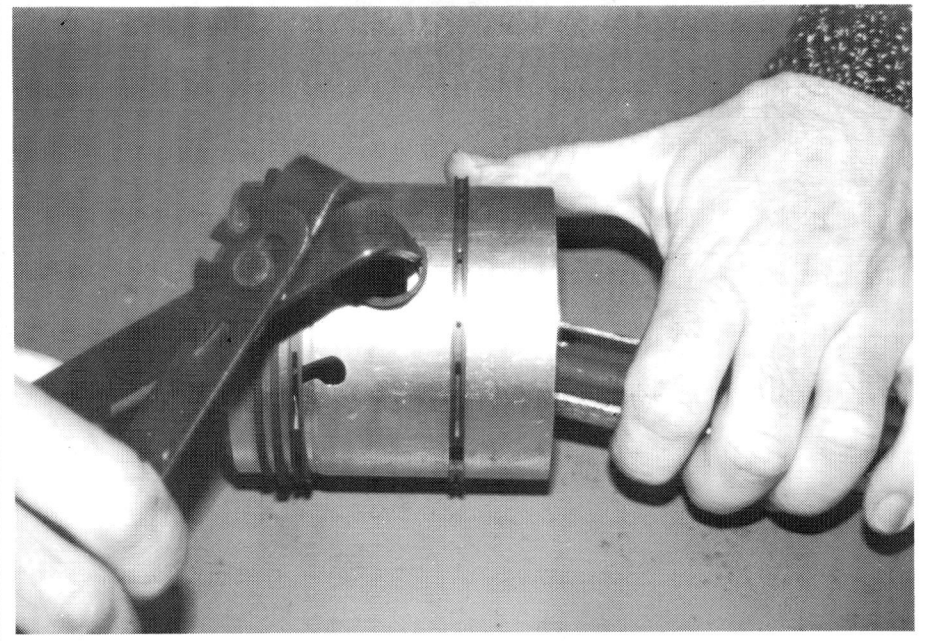

Using good quality circlip pliers, install the circlips to hold the pin in place. Make sure it is fully home – if it should work loose disaster would follow. Although the connecting rod should move freely on the gudgeon pin, there should not be any excessive play.

It is worth renewing the big end nuts and bolts, particularly if, like these, the nuts are castellated and split pins are fitted. Use the new type self-locking nuts.

Remove the sump baffle assembly and thoroughly clean. This is very important as there could be unwelcome sludge ready to run amuck in our spotless engine.

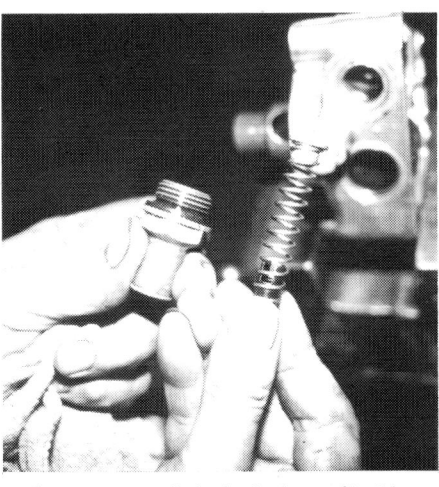

The pressure relief valve is situated inside the by-pass outlet. Both the valve and the spring should be replaced.

Fit new crankshaft bungs and spot punch at the ends for a permanent fix.

**Next Issue
We start to assemble the main unit.**

"OUR ENGINES SPEAK FOR THEMSELVES"

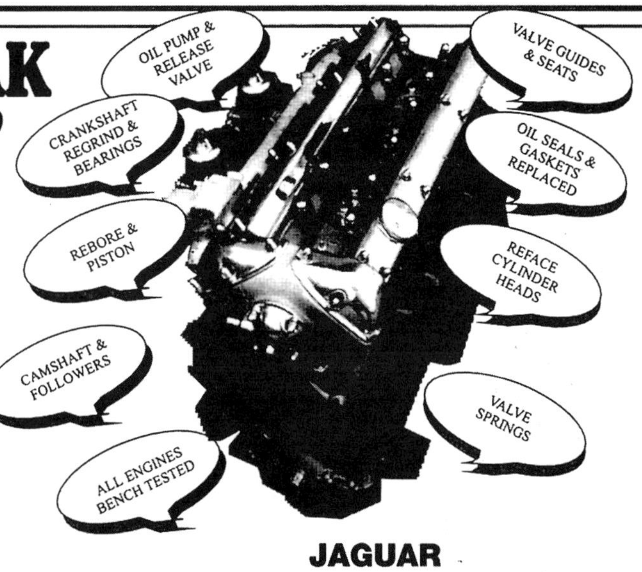

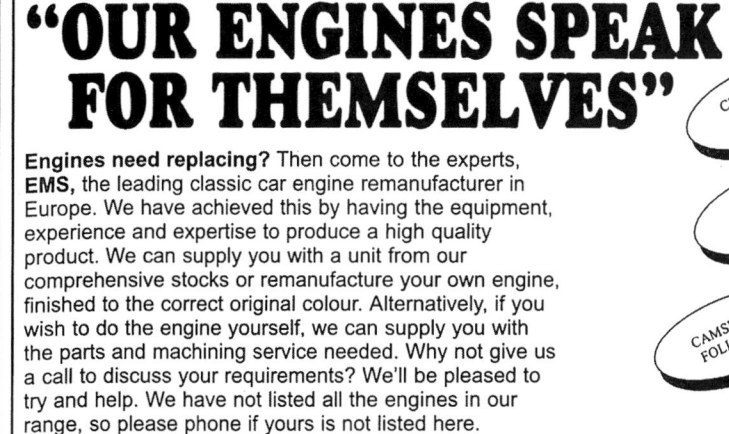

Engines need replacing? Then come to the experts, **EMS,** the leading classic car engine remanufacturer in Europe. We have achieved this by having the equipment, experience and expertise to produce a high quality product. We can supply you with a unit from our comprehensive stocks or remanufacture your own engine, finished to the correct original colour. Alternatively, if you wish to do the engine yourself, we can supply you with the parts and machining service needed. Why not give us a call to discuss your requirements? We'll be pleased to try and help. We have not listed all the engines in our range, so please phone if yours is not listed here.

FACILITIES AVAILABLE

- Engine Fitting Service
- Surface Grinding
- Cylinder Reboring
- Crankshaft Regrinding
- Pressure Testing
- Electronic Crack Detection
- Chemical Cleaning & Shot Blasting
- Drilling, Milling & Turning
- Aluminium Welding
- Full Balancing Service

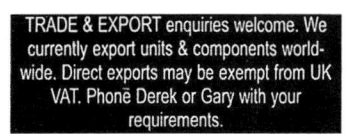

BRITISH MOTOR HERITAGE APPROVED

DETAILED TECHNICAL RECORDS KEPT ON ALL ENGINES

Engine labelled: OIL PUMP & RELEASE VALVE, VALVE GUIDES & SEATS, CRANKSHAFT REGRIND & BEARINGS, OIL SEALS & GASKETS REPLACED, REBORE & PISTON, REFACE CYLINDER HEADS, CAMSHAFT & FOLLOWERS, VALVE SPRINGS, ALL ENGINES BENCH TESTED

JAGUAR

Our rebuilds are based on age and capacity of unit. Prices range from between £1,978 and £2,375 fully rebuilt. *Example:* XJ6 Series 1, 2 or 3 from £2,000 (exchange). For lead-free engines, please add £230. All engines carry our 12-month unlimited mileage F.E.R. approved warranty with a full spec. sheet available on request.

TRADE & EXPORT enquiries welcome. We currently export units & components world-wide. Direct exports may be exempt from UK VAT. Phone Derek or Gary with your requirements.

SPECIAL Nationwide collection or delivery service. Mainland UK only £18 per unit.

Federation of Engine Re-Manufacturers

EMS

ENGINE MACHINING SERVICES LTD
KIRKHAM WORKS,
CENTRAL AVENUE,
WORKSOP, NOTTS, ENGLAND S80 1EN
TEL: 01909 482649. FAX: 01909 479033
MOBILE: 0836 323003

ALL PRICES INCLUDE UK VAT

Access / VISA

Before you renew your present policy or when buying your latest *Jaguar.*

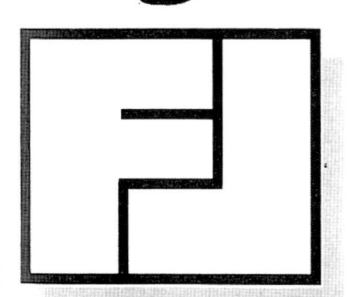

JAGUAR ENTHUSIASTS CLUB

Write, Phone, or Fax for a Quotation

JAGUAR ENTHUSIASTS INSURANCE HOTLINE

TEL: 0121 561 4196   FAX: 0121 559 9203

Footman James & Co. Ltd. Waterfall Lane, Cradley Heath, West Midlands. B64 6PU.   Tel: 0121 561 4196   Fax: 0121 559 9203

MARKET LEADERS for Over a Decade

The Enthusiasts Insurance Broker

The XK ENGINE

PART 7 – ASSEMBLY

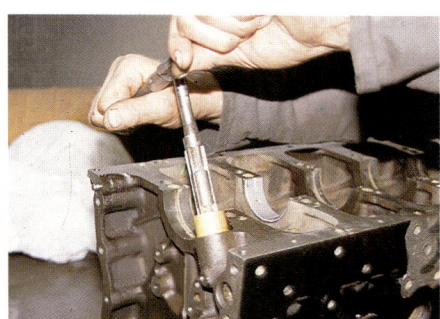

Fit a new distributor drive shaft bush by pressing into its bore. It should then be reamered to a diameter of ¾in +0.0005in or – 0.00025in. Or use the old shaft and adjust receiver to suit.

Press the new seals into the rear oil return thread covers. Do not cut off any excess but continue to push the seals in. Using a socket bar will eventually achieve the result.

The money's spent and the machining done. All that background work has been leading up to this, the final moment of assembly. More than ever before it is important to have a clean working area with everything at hand. It is worth borrowing or hiring an engine stand to support the block or, if this is not possible, then place a wooden block at each end of the engine to keep it away from the bench or floor.

Before assembly begins it is advisable to go right through the engine and its components to check that every thread is clear and that all surfaces are free from burrs. A tap and die set will ease the threads but do make sure you use the *correct* size and *never* force anything.

All new components should be thoroughly degreased; we have already stressed how important it is that all used components should be thoroughly cleaned, and that when it comes to block and crank this can really be achieved only through dipping at a professional engine shop.

Arrange the engine block so that the crankcase faces up (ie. upside-down) and check that the main bearing shells are clean and free of burrs.

Secure the top oil return cover, with a little Hylomar applied to the inner face. Note that the bottom cover bolts are of different sizes – the shortest is fitted in the middle. Fit the rear main bearing cap (without shells) and tighten to a torque of 83lb ft.

Fit the bottom oil return cover and then coat both seals with grease.

Using a sizing tool (Churchill no. J17, available from the Jaguar Enthusiast Club), press and turn until it is fully home. Withdraw by turning in the reverse direction, slightly pulling. Remove the bottom half cover and bearing cap. It is possible to achieve an effective fit by the using the socket bar alone but remember that this seal is at the very rearmost of the engine; the choice is yours.

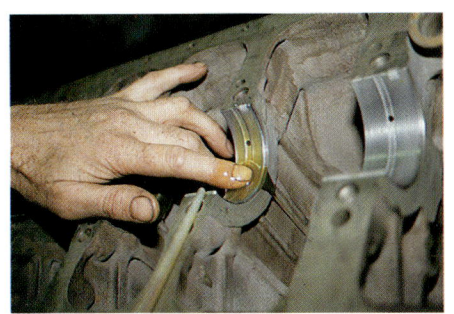

Push the main bearing shells into position ensuring that the locating notch is firmly home. Liberally coat with oil.

Using either a dial gauge on the end of the crankshaft or feeler gauges at the centre main, check for end-float, which should be 0.004-0.006in. If the measurement exceeds this then the oversize thrust washers will be needed.

Push remaining bearing shells into caps and coat with oil.

The centre bearing cap is fitted with thrust washers (available in two sizes, std and 0.004in oversize). Using the standard size, fit to the block ensuring that the number on the cap matches the corresponding number on the block. Tighten to a torque of 83lb ft.

When all bearings are in place the crankshaft can be lifted into the block. Liberally coat all bearing surfaces with oil.

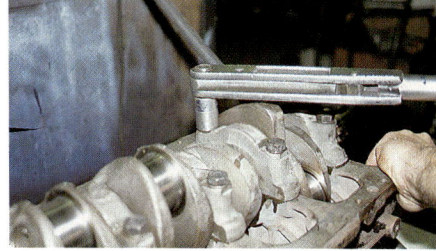

Fit the rest of the main bearing caps, paying special attention to the identification numbers on the caps and the block. Tighten to a torque of 83lb ft. Don't forget to replace the oil pump pipe stabilising brackets.

Turn the crankshaft to make sure it spins freely without snagging or resistance. Two bolts screwed into the end of the crankshaft with a lever inserted between them is a most convenient method. If you are satisfied, turn up the tab washers under the bolts.

Replace the oil gallery plugs. The one at the front (brass on early XK engines but interchangeable with later types) should be smeared with Hylomar or similar compound. The rear and side should be fitted with new copper washers.

Replace the core plugs; early engines use the dished variety and are struck in the centre to spread into the recess. Later type are cupped. A small smear of Loctite around the edge aids the sealing quality. Use a broad based drift to drive the plug home. On 4.2 blocks with long studs it is worth leaving the core plugs until the studs have been fitted as threads are just accessible through the holes.

Don't make this mistake – the piston ring gaps must be staggered to hold compression.

Thoroughly clean out the bores and crankshaft journals before coating everything with oil. With the crank-pin at bottom dead centre, select the correct numbered rod with its new piston, suitably oiled, and place in the bore. Remember that Jaguar number the bores with no. 1 at the rear. A piece of plastic tubing on the big end bolts will protect the crank-pin from any chance damage.

Using a good quality ring compressor, compress the rings and, using a soft weight such as the wooden handle of a club hammer, drive the piston into the bore.

Put the big-end shells into the con-rod and cap, thoroughly oil and fit the cap with the corresponding numbers matching. They should be on the exhaust side.

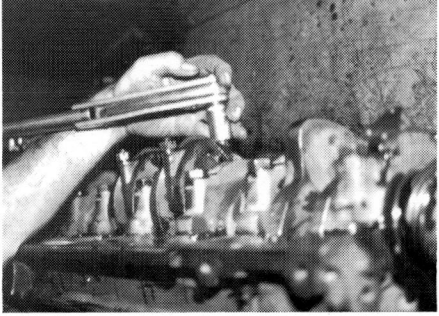

Tighten to a torque of 37lb ft and repeat operations 18, 19 & 20 for all pistons. When complete, spin the engine a few turns to check that everything is free.

With no.6 piston at top dead centre, fit the distributor drive-shaft in the bearing with the larger portion of the offset drive to no. 6 piston.

Fit the thrust washer, Woodruff key and then slide the helical gear into position. Tighten securing nut but be careful to support the square end of the shaft otherwise damage may occur to the bronze gear. There should be end-float of 0.004-0.006in. Knock the tab over when satisfied.

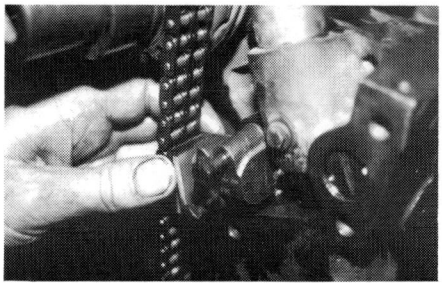

Replace the conical filter in the oil feed hole and, using shims if necessary, fit the timing chain adjuster so that the chain runs centrally on the slipper block. Tighten two bolts and knock over tab washers. Earlier adjusters were released by an Allen key in the rear but later replacements have a slip of cardboard locking the plunger. DO NOT RELEASE THE PLUNGER UNTIL THE CHAINS ARE IN POSITION.

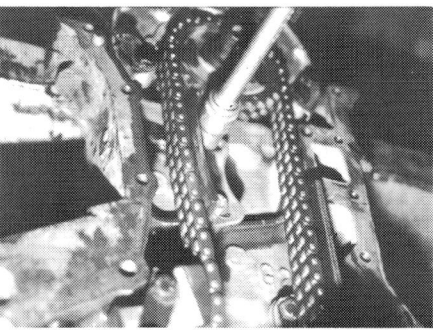

On 4.2/2.8 and later 3.4 engines, there is an intermediate damper secured to the block. This should be in light contact with the chain when the slipper to tensioner body gap is ⅛in.

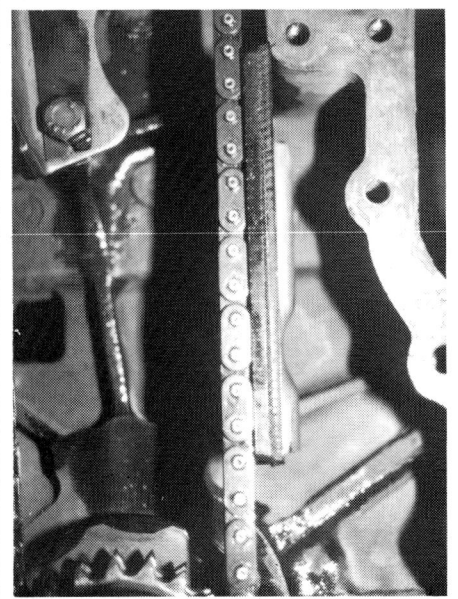

The lower damper (4.2 and subsequent engines) should be in light contact with the chain. These bolts are inaccessible from above and a cranked spanner will be needed.

Secure the lower chain damper to the block. Taking the timing chain sub-assembly, loop the bottom chain over the crankshaft sprocket and fit the mounting brackets, with the two dampers sandwiched, to the block. On 3.4/3.8 (see later for 4.2 engine) engines fit the intermediate damper to the bottom of the rear of the mounting bracket with slotted setscrews. Fully tighten the four long bolts first and two setscrews after.

Position the front oil seal midway along the crankshaft nose. The kit usually comes with a new seal track to complete a perfect seal. Polishing the crank nose with special engineer's paper will take off any minor burrs.

Smear Hylomar into the oil seal groove and offer the timing chain cover on to the oil seal and push home. It is worth applying Hylomar lightly to the gaskets to keep them in place. Make sure the right length bolts are used for the right holes.

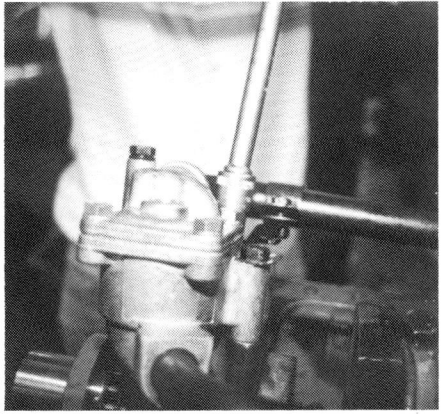

Place the oil pump coupling on to the distributor drive and lightly bolt the oil pump into position. Ease the delivery and suction pipe into their respective ports.

Using a new gasket, bolt the delivery pipe to the block and fix the securing brackets. It will be necessary to remove a main bearing bolt to facilitate this operation if not done earlier. Finally, tighten oil pump.

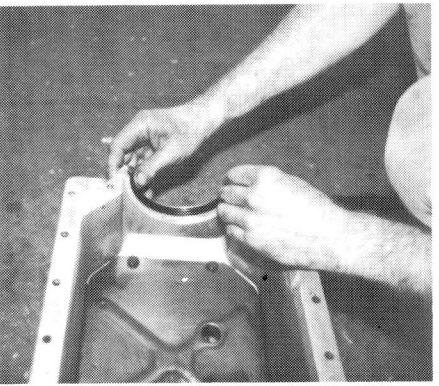

Smear Hylomar into the oil seal groove and push the rear oil seal into the sump. Lay the sump gasket into position on the block with a light application of Hylomar.

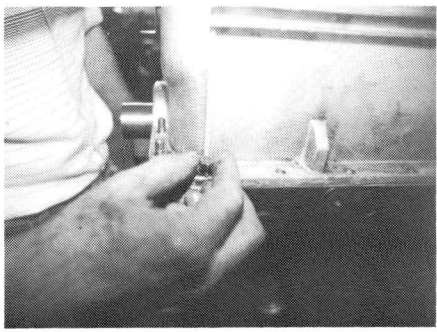

Lower the sump on to the block and push home. Gradually tighten the retaining bolts noting that the shortest is fitted as shown.

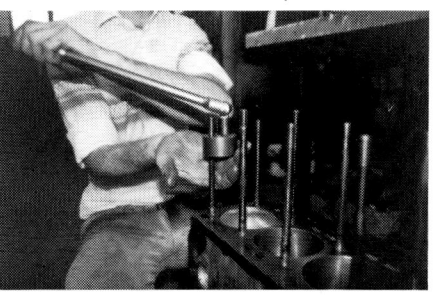

Stand the engine on its sump and, using a good quality stud extractor or two nuts locked together, replace the cylinder head studs. Ideally use new studs, especially on the 4.2 engine. Note that a dowelled stud is fitted on the exhaust side second row from the front. Refit the four small studs at the front of the cylinder head.

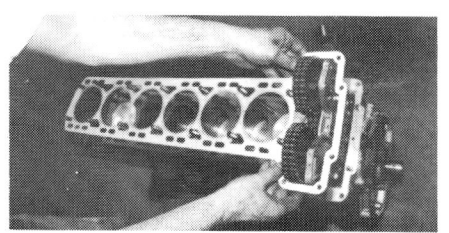

Make sure that the timing chains are on the most innermost part of the block and lock them in position (stretch an elastic band across the two gears). Lay the head gasket in place. Check that the right one is being used as there is a bewildering variety.

With the engine set at TDC and the camshafts set in the same position (use Jaguar's valve timing gauge), engage the help of a burly assistant and lower the cylinder head on to the block.

Fit engine lifter and spark plug lead carrier if applicable, two plain washers to the front two studs and 'D' washers on those remaining. Tighten the 14 chrome dome nuts to a torque of 54lb ft in the sequence shown. Do not forget the six nuts at the front, and the copper washers where in contact with aluminium.

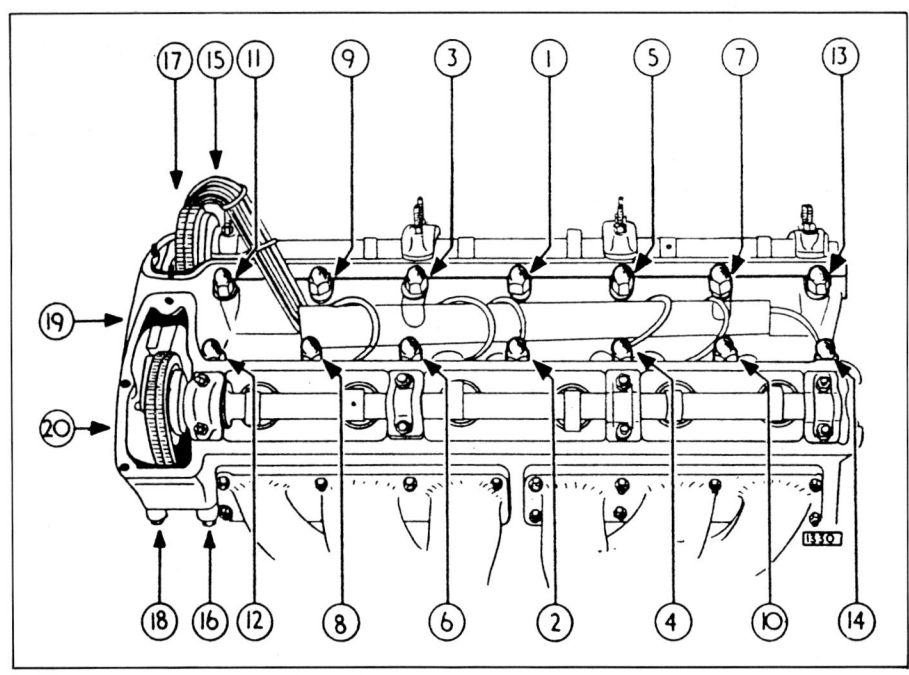

Tightening sequence for the cylinder head nuts.

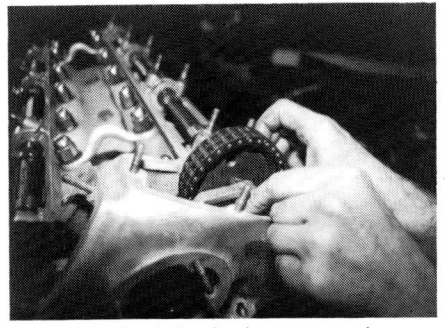

Double check that both pistons and camshafts are at TDC. Ease the timing chain and sprocket over each camshaft and align the camshaft adjuster plate bolt holes with the corresponding thread holes in the camshaft. Rotate through 180° if the holes do not line up. Push the plate into the splines on the sprocket and put the circlips in place; this a very fiddly operation. Fit a bolt in the visible hole and 'nip' home. Turn the engine until the other hole is accessible and fit the remaining bolt. Fully tighten all bolts.

Lock all bolts in position with locking wire.

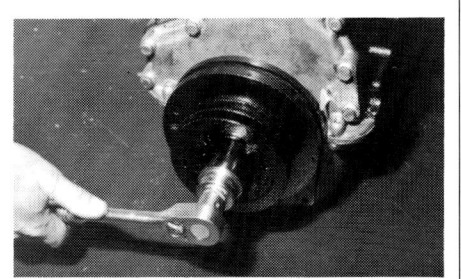

Fit the Woodruff key and then the split cone to the nose of the crankshaft and, with the pulley attached to the crankshaft damper, offer up the assembly. Fully tighten the large centre bolt and fit the locking tab. Re-check that the engine is at TDC and fit the timing marker to 0 deg. on the timing scale on the pulley.

If manual transmission is fitted, replace the spigot bush in the end of the crankshaft. After checking that the crankshaft and flywheel are free from burrs, offer the flywheel into position with the balance mark 'B' at approximately BDC. Tap the two mushroom dowels into position and, using a new locking tab plate, fit the securing bolts and tighten to a torque of 67lb ft.

Slacken the locknut securing the serrated plate on the top timing chain adjuster, located through the breather aperture. Using a timing chain adjuster tool No. J.2 (available from the Jaguar Enthusiast Club), press down the locking plunger and insert the two prongs into the two holes in the serrated plate. Turn in an anti-clockwise direction until there is just a little give on the chain. Re-check the valve timing and, if necessary, readjust.

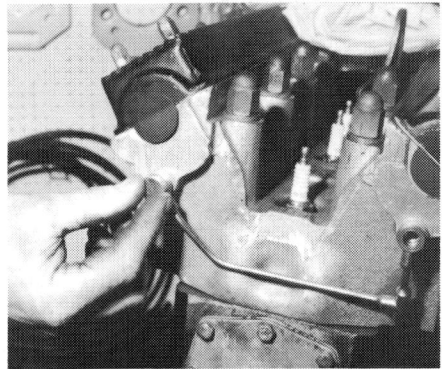

Using new copper washers, fit the oil feed pipe from the block to each camshaft bank. If the threads in the head look a little suspect, the Jaguar Enthusiast Club offer extra long bolts to compensate for this.

The oil filter housing can now be fitted to the block using a new gasket. Fit a new bypass hose and clips.

All that is now required is to fit the camshaft covers, front breather cover, distributor and ancillaries and the engine is ready to go back in the car. Note that there is a torque setting for the cam cover domed nuts – 15lb ft.

NEXT ISSUE
Cylinder head identification, and refitting the engine.

THE SIGNAL

THE SOLUTION

With over 50 years experience supplying piston rings to motor, aero and marine industries Cords reputation speaks for itself.

If you are rebuilding, or just reconditioning, a Jaguar, Cords will almost certainly have the replacement piston ring kit for your engine.

To help complete the task, Cords supply glaze busters, ring compressors and bedding in oil.

For further information and specific vehicle enquiries contact Paul on:
Tel. 0181 998 9923 Fax. 0181 998 1214

CORDS ENGINEERED piston rings

CORDS PISTON RING COMPANY
(A wholly owned subsidiary of Bluecol Brands Ltd)

**8 AINTREE ROAD,
PERIVALE, GREENFORD,
MIDDLESEX UB6 7LA.**

CORDS
AT THE HEART OF A BETTER ENGINE

STOP
STOP
STOP

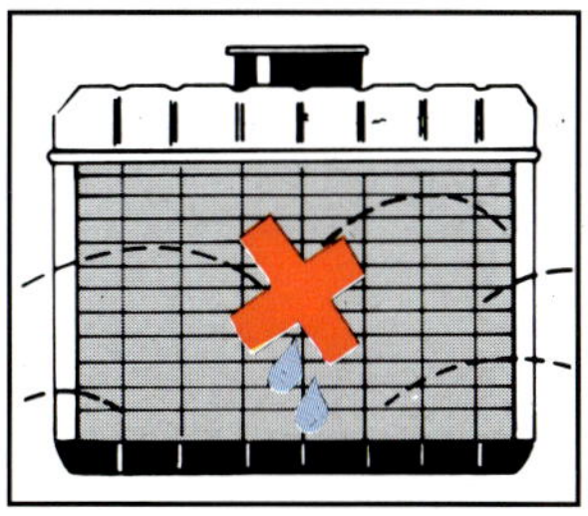

Bars Seal. The cooling system sealer. Acts from the start to stop leaks and prevent seepage – permanently. Used by major motor manufacturers in many new engines. Powerful and effective in older vehicles.

Bars Seal. Crumble a pellet into your radiator. And seal any leaks within minutes of starting your engine. Harmless in contact with rubber, plastic, aluminium and all other metals. Totally compatible with antifreeze.

Bars Seal. Stop believing there's anybetter way to stop leaks. And stop by at your local Bars stockist today.

BARS MOTOR PRODUCTS LTD
tel 0181 998 9923
fax 0181 998 1214

A wholly owned subsidiary of
BLUECOL BRANDS LTD

Prevention
is better than cure

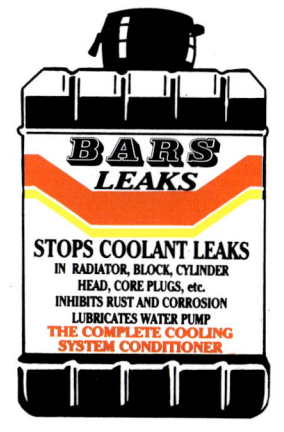

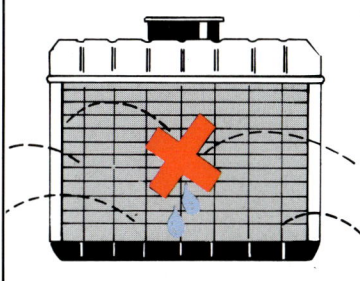

Bars Leaks. The best protection for any cooling system. Acts to prevent coolant leaks right from the start. Stops sludge and limescale formation. And lubricates water pumps.

Bars Leaks. Its unique formula circulates with the coolant in many new vehicles. And is equally powerful and effective in older systems.

Bars Leaks. Don't wait until cooling system leaks leave you by the roadside waiting for the cure. Prevent that possibility today. Call into your nearest Bars Leaks stockist now.

BARS MOTOR PRODUCTS LTD
tel 0181 998 9923
fax 0181 998 1214

A wholly owned subsidiary of
BLUECOL BRANDS LTD

The XK ENGINE

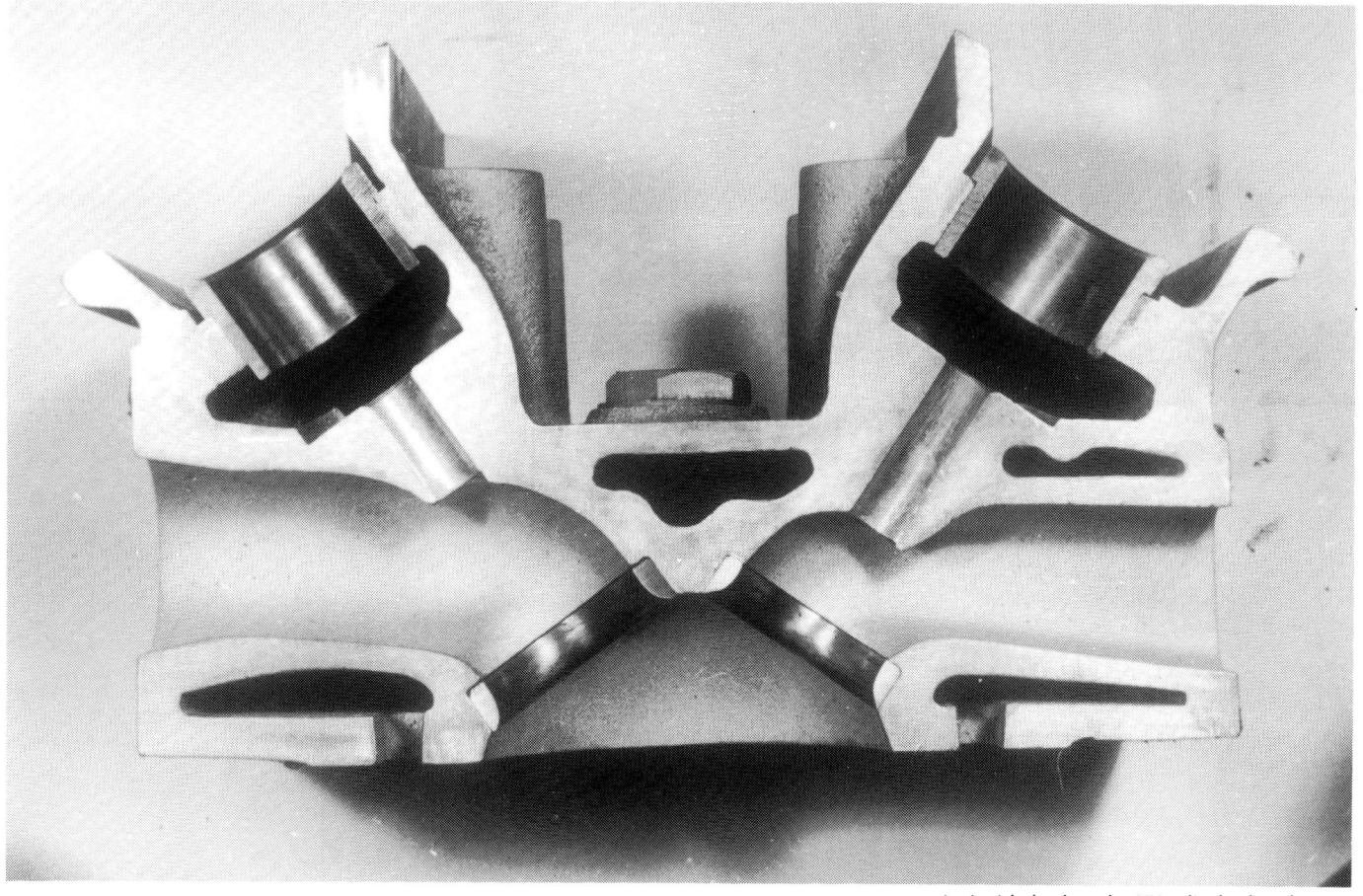

An inside look at the XK cylinder head.

Head'n for trouble?

**PART 8 –
Jim Patten sorts out your cylinder head dilemmas**

"I've bought this 'E' type see, an' it ain't go no 'ead on the ingine. Bloke down the pub reckons the one off 'is ol' Mk X will do, but I dunno."

Well, bloke could well be right. Jaguar engines are blessed with a certain degree of interchangeability but the area can be a minefield. With caution, however, it is possible to fit parts from the last Series 3 XJ6 to even the very earliest XK 120. But here we are concentrating on cylinder heads – how you identify them and how they can be swapped around.

The head stud pattern follows the same format throughout the range so, theoretically, a big-valve Series 3 XJ head will physically fit a Mk VII. However, certain rules apply with the original 'A' type head as fitted to the XK 120 and Mk VII laying down the basic pattern of things to follow. Larger valves and different valve angles on the 'C' type soon found their way (slightly modified) into production on what would become the 'B' type head. All of these heads are interchangeable using the standard inlet manifold arrangements. Then along comes the XK 150S with its increased power, part of which can be attributed to the new 'straight port' cylinder head. Again this head will fit all engines but, due to the revised porting arrangements, only with the correct inlet manifold. Both 'E' type and Mk X use the straight port head but, while their inlet manifolds appear to be identical, scrutiny reveals quite different arrangements.

First, the 'E' has manual choke operation and secondly its manifold slopes downward to clear the bonnet. 3.8 straight ports used a three-part inlet manifold with either a threaded bung or a core plug to block the water passage. The manifold had an external water passage. 4.2 heads (all straight port of course) incorporated the head passage in the manifold by using a one-piece casting, so no bungs in the head. Then the 420 joined the range, giving at last a straight port head with just two carburettors – an ideal package for an easy increase in bhp for a Mk 2 which hasn't room (without cutting) for three carbs. Both 240 and 340 used the same head but with smaller carburettors.

Complications set in with the XJ and the last of the 'E' type/420/420G cars. Jaguar set about a serious redesign of the engine block, tying the crankcase to the cylinder head with much longer studs. In addition to this a 'coffin' end was added to the block giving an extra couple of water holes. Fine for the XJ but it does make life a little more difficult on our interchangeability trail.

There really is little to be gained by fitting an XJ head to an earlier model as opposed to an earlier straight port head, and sourcing a 420 type will avoid various problems. Unless, that is, you use the Series 3 head and then only the 4.2 injection model has the advantage of larger inlet valves. It's worth noting that Series 3 injection heads are still available from the factory and, let's face it, there's

The XK ENGINE

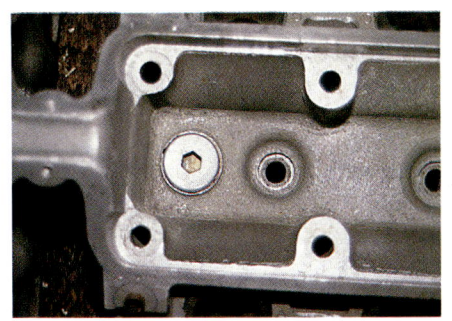

'A' type were not painted, being left natural aluminium. (production 'C' type is painted red with a raised 'C' cast into the spark-plug well).

'B' type 2.4 & 3.4 were painted a duck egg green. Ken Jenkins Jaguar Specialist (Tel: 0909 732219), claims to have the correct colour in stock.

'B' type – 3.8 painted a metallescent blue; early heads had a chamfer around the combustion chambers for piston clearance. This was altered in 1962 when the pistons were modified. The correct colour is very difficult to determine as it seems Jaguar did not keep to the same shade. Picture illustrates original factory finish.

Very early 'E' types and Mk Xs, and some XK 150S cars, were painted a plain dark orange/yellow. An original, untouched XK 150S is illustrated.

nothing like working with new parts.

This head will fit the block as normal, it's just that the extra cooling holes overhang the block edge. To overcome this problem, simply tap a thread and screw in a bung using a quality thread sealant as supplied by Loctite or similar. The obvious advantage of using a pre-XJ straight port head means that this fiddly operation can be avoided. Whatever head you choose, always use the appropriate gasket.

It should be noted that, although the cylinder bore positions were altered when the 4.2 block arrived, Jaguar never changed the basic cylinder head casting and a slight misalignment always remained between combustion chamber and bore on 4.2-litre engines!

Camshafts

A word or two about camshafts. There are only three basic types of production camshaft. 1/ The low-lift as fitted to Mk VII, XK 120 and early 2.4. 2/ The standard 3/8in lift used by far the vast majority of 1950s/60s Jaguars. 3/ In 1969, profiles were changed to give quieter running but the lift was not changed. Make sure that, if these later cams are fitted, the correct valve clearances are used. Later cams are also four-bolt fixing with injection inlet cams different again. Volume 4 No.1 gives more details in the cylinder head overhaul section.

So you think, why bother with any of this at all? Well, I suppose it could have something to do with our basic lust for power. Or it could be that the old head has simply had it and, unless you have a fetish for matching numbers, then swapping heads can make good economic sense. In ascending order of efficiency the heads run A, B/C, straight port, and

47

Straight port heads are instantly identified externally by the repositioning of the spark plugs, the bung appearing between the first two plugs. XK150S, 3.8 'E' type and 3.8 Mk X are painted gold.

Series 3 injection at the top. Competition heads are another matter but the production 'D' type used the production (ie. XK 140 optional) 'C' type head casting modified to take 1⅞in inlet valves fitted.

Carburation

Mix'n match runs a little foul in some areas of carburation. An XK 120 in rhd form cannot run with three SUs as the steering column precludes this. Similarly, the same arrangement on the Mk 1 & 2 is obstructed by the clutch reservoir. The 'S' type is a little easier with its remote container. By far the easiest way is to use the head and twin carburettors from a 420. An XK 120 needs care, as even when fitting a 'B' type head and manifold the steering column is still perilously close.

The best thing for early 2.4 cars is to junk Solex and favour SU, just as Jaguar did later. Don't forget that if you abandon the air cleaner the engine effectively runs weaker so a change of carburettor needles will be needed.

On a more tedious note, don't forget to advise your insurance company on any major change relating to performance. They get dreadfully upset at secrets.

NEXT ISSUE

– final tuning on the rolling road

'E' type Series 2, 240/340, 420G, 420 and XJ6 – and late 'B' type – are painted silver. Series 2 'E' type, late 420G and XJ6 heads were slightly longer to accommodate two extra water passage drillings.

3.8 straight port heads used a bung on the inlet side of the water passage, removed on the later heads where the water passage continues into the manifold.

The cylinder head was changed from the following numbers to accommodate two extra water holes. 420, 7D58882; 420G, 7F11173; 'E' type, 7R1915 & 7R35389; XJ6, 7L1176. Note that the head is slightly longer.

VSE

JAGUAR

Specialist engine reconditioners and parts suppliers

20 YEARS SERVICE TO THE TRADE

REALISTIC PRICES TO THE ENTHUSIAST

WARNING

DO NOT REBUILD YOUR JAGUAR ENGINE...

...without first obtaining your FREE engine parts and information book from **VSE** *(trade and export welcome)*

VSE stock a full range of engine components for Jaguar 6 cylinder engines at substantial discounts including; pistons, rings, bearings, gaskets, seals, camshafts, conrods, bushes, ring gears, valves, guides, cam followers etc, etc., plus all the important lock tags, bolts etc.

VSE also offer a full range of machining services including; rebores, crack testing, crack repairs and crack grinding, balancing, valve guide and seat fitting, surfacing etc, etc, all at realistic prices.

VSE also stock a range of exchange, cylinder heads, engines, cranks, rods, etc.

VSE can naturally rebuild your own engine, tuning, conversions, lead free, dyno testing. In fact, for anything to do with engines contact **VSE**.

VSE
Llanbister, Llandrindod Wells, Powys
LD1 6TL

Tel: 01597 840 308 / 0831 280157
Fax: 01597 840 661

JAG SHOP

The Jag Shop is a well established Jaguar parts supplier covering 60's, 70's to 80's, 90's models. We have been established since 1976 and have gained a worldwide reputation for our friendly and knowledgeable service. A comprehensive inventory of all Jaguar parts both fast and slow moving are kept in stock. We have workshop and engine rebuilding facilities available with competent technical staff to help you.

Next time you have an enquiry or need engine parts please contact one of our staff who will be pleased to help you. Free price lists are available on request.

JAG SHOP

JAG SHOP • 303 Goldhawk Road, London W12 8EZ • Tel: 0181-748 7824 • Fax: 0181 563 0101

Jaguar World, November/December 1992

The XK ENGINE

The Tune-up

Part 9 – Jim Patten takes a drive to nowhere on the rolling road

I t could be an engine built by a grand maestro but if the distributor is wrongly timed or the carburettors are out then all his meticulous work would have been in vain.

In a perfect world, every rebuilt engine would be put on a test bed and fired up before it goes anywhere near a car. Thoroughbred Cars and other quality engine builders have their own purpose-built test cells where all the adjustments are made prior to fitting. Of course, you won't have these facilities at home but some companies will carry out the work for you. However, carting engines about is a laborious job and a good second-best is to fine-tune a 'new' engine in the car on a rolling road. We went along to see Roy Martin and Nick Smith at Swallow Engineering with our 'competetion coupe' and to see the full works applied to a 3.8 'E' type.

Before any work commences, Roy checked the engine oil and water levels. As the car will be run effectively at high speed, the tyre pressures are also checked.

This piece of kit is the brains of the operation. Its first job is to check the current draw on the starter motor and that the coil is in good order. Then, with the engine running, the charging system is assessed; at 13.5 volts, we were OK.

An array of switches controls the various functions; next to be checked was the consistency of firing. The screen shows each of the six cylinders firing. Each 'tail' should be equal; if not, then it could be that the distributor cam is worn and each cylinder has slightly different timing. The timing is checked through the rev range to make sure that the rest of the distributor is functioning correctly.

51

Clips everywhere. Here the firing voltage is checked and, by cutting each cylinder in turn, the reaction is demonstrated by showing a power loss. Equality is the ideal but close will do.

The carburettors should be inspected to make sure that the piston slides freely and that the dashpots are correctly filled with the right light oil. Our coupe was a little soiled (left) but a simple clean brought things back to new (right).

Continued over

Spark plugs are checked. We use a standard feeler gauge but here Roy has a Snap-On gauge/adjuster – it's needed on a 12-cylinder Jaguar!

The next routine adjustment is the points gap.

With the plugs out, the cylinder compressions are measured. Optimum figures are different from engine to engine depending on compression ratio but all should be within 10% of each other.

Now Roy prepares to recheck the ignition timing on the car using a strobe light. It is useful to put a dab of white paint on the appropriate line on the crankshaft damper scale and another on the pointer. On the 'E' type this is situated at the bottom.

Jaguar World, November/December 1992

The XK ENGINE

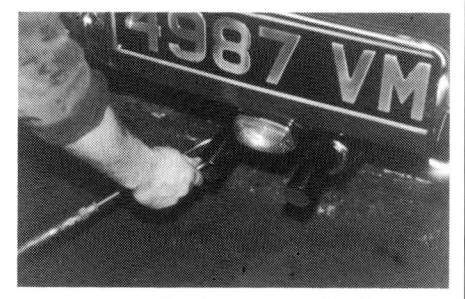

Here the Co level is measured and accurate mixture adjustments made.

The final adjustment will be to equalise the carburettors and adjust the slow running. There is the myth of the ol' boy with a roll-up, short back and sides and 2in braces doing this job by ear and a piece of hose. He's probably very good but I'll take the flowmeter shown here.

When everything is done the car is then run on the rolling road to show the power output at the back wheels. This is usually before and after tuning to show the benefits obtained.

If you want to test your engine prior to installation, companies such as E.D.S. of South Ockendon, Essex, will run your engine on a dynamometer and return it fully tuned and ready to fit complete with full computer printout showing bhp and torque figures along with all the settings. This costs around £200 plus VAT.

Swallow Engineering will carry out a basic rolling road tune on a six-cylinder Jaguar for £55 plus VAT and any parts. Obviously 12 cylinders cost more.

Our thanks go to Roy Martin and Nick Smith at Swallow Engineering, 6 Gibcracks, Basildon, Essex (Tel: 0268 558418, fax 0268 555743), and Alex Macfadzean of EDS Engine and Dynamometer Services Ltd, Arisdale House, Arisdale Avenue, South Ockendon, Essex RM15 5AS (Tel: 0708 857108).

NEXT ISSUE
A brief look at early and competition engines.

THE INDEPENDENT ALTERNATIVE

XJ40, XJS & SERIES 3 PARTS SPECIALISTS

We are specialist parts suppliers for all late model Jaguars.
Friendly service at competitive prices

PARTS EXPRESS HOTLINE
TEL 01325 332505 FAX 01325 332405

EXPORT WELCOME

ASK US ABOUT SERVICE KITS – GENUINE FILTERS – SERVICE ITEMS – SUPERB SAVINGS

CLASSIC COMPONENTS, TOWER HOUSE, TEESSIDE INTERNATIONAL AIRPORT, DARLINGTON DL2 1PD

OVERNIGHT DELIVERY UP TO 6pm FOR NEXT DAY

Ken Jenkins
M.I. Diag. E.. C&G Tech Cert.

Spares for models 1950 onwards • Classic Jaguar Sales
Special Trade Prices on many items

Unit 4, 2 High Road, Carlton-in-Lindrick, Worksop, Notts S81 9ED
Tel: 01909 732219 • Fax: 01909 731032

Jaguar World, March/April 1993

Jaguar goes Independent

In Part One of our new series, Jim Patten takes an initial look at Jaguar's first and immensely successful independent rear suspension.

IRS – Independent Rear Suspension

Can the overhaul be done at home? Will I need special tools? Should I trust myself or give it to an expert? Read on for the answers – you'll be surprised.

Jaguar paid a brief visit to independent rear suspension systems when they were involved in war work, developing a small lightweight Jeep. But it was at the cessation of hostilities that Bob Knight investigated IRS seriously and looked at a system patented (but never used) by Georges Roesch of Clement Talbot with a view to adopting it for their forthcoming range. It was accepted in general terms but left on the back boiler until the mid-fifties, when it was agreed to incorporate an IRS system into their new series of cars ('E' type and Mk X).

Although they played with a de Dion for the 1956 'D' type, it was on the 'E' type prototype, E1A, that the first production-type IRS was used. Initially the differential unit was bolted directly to the main body unit and, while it worked well enough, too much vibration was transmitted into the car.

Roesch sought to use the driveshaft in place of the lower wishbone with a solid upper link swivelled at the inner and outer points. Tests proved this to be extremely effective and Jaguar followed the formula but with the driveshaft uppermost. Disc brakes were mounted inboard with the driveshaft running through two universal joints to a large aluminium hub carrier. A substantial tubular lower member, yoked at each end, was coupled to the differential and the bearing housing. Two coil over damper units were hung each side.

After trying various strengthening points on the chassis it was decided to use a separate 'cage' to enclose the whole unit. This would be secured to longitudinal members by four Metalastic frame mounts. For positive for and aft location a radius arm was later added, secured to the lower wishbone and the rear of the floorpan.

It was this guise that made the production line and in 1961 both the 'E' type and Mk X became the first production Jaguars to feature fully independent suspension on all four wheels. It's still going strong today in the XJS and has only just been phased out with the demise of the Series 3 XJ12 and limousine; XJ40 uses another system. As with so many components, regular maintenance makes this an extremely reliable item. Special and replica builders all over the world find the conveniently-packaged unit an absolute marvel and its popularity for many projects continues.

The sad fact is that, tucked away at the rear of the car, the IRS is often not only forgotten but seriously neglected. Until the autumn of '92 the rear suspension was not even liable to checks during the MoT test, a situation thankfully now remedied. Older Jaguars also had an

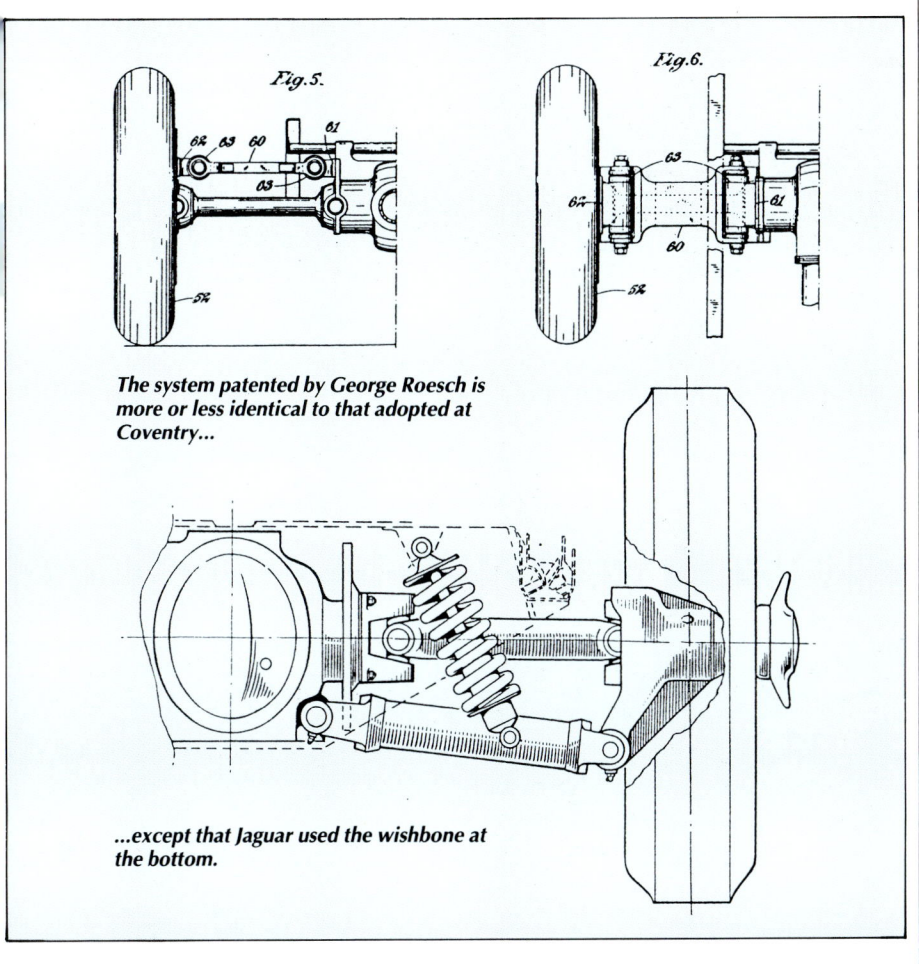

The system patented by George Roesch is more or less identical to that adopted at Coventry...

...except that Jaguar used the wishbone at the bottom.

Using a large trolley jack with a stout block to spread the load, the axle can be lowered and removed from the car.

undeserved reputation for poor handbrake operation. Given that the brake discs are in good condition and the pads are not worn, then the handbrake should be effective. The trouble is, of course, that all this mechanism is hidden and a little tricky to get at but can be maintained with the right approach.

The first cars equipped with this suspension are now over 30 years old, while even early Series 3 XJ6s are knocking on for 12 years. The chances are, especially if the car feels horrible to drive, it hasn't been maintained properly and while odd, unevenly worn tyres can affect the handling, it's probable that there is a worn-out rear end with shot fulcrum shaft bearings, drive shaft UJs and rear frame mounts. With all the faults corrected the car will be transformed, so it's a job very well worth doing.

A good whack with a hammer to the centre of the Metalastic bush should free the radius arm from its mount.

55

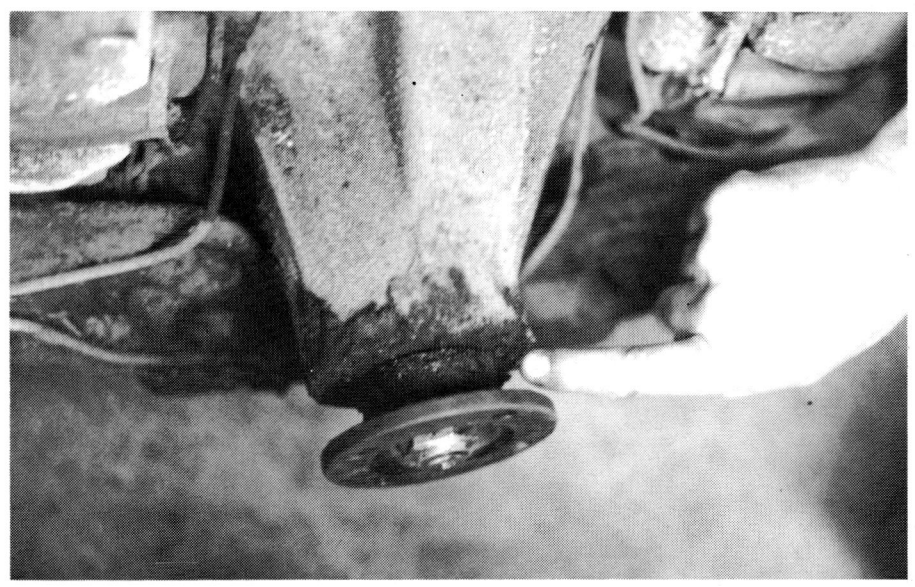

With the axle on the floor, a look around the propshaft flange shows that the oil seal has been leaking for some time. The drive shaft output flanges are also moist so oil will have been issuing forth from here as well. The obvious danger is oil spraying on the inboard disc brakes.

This radius arm rubber is about to part from its seating. Had it remained on the car then the acceleration/de-acceleration would have been very rocky indeed.

Rot is present on this 'E' type but luckily the radius arm mount is still in position. In severe cases, this whole section can just fall to the ground after a good burst of acceleration.

Axle removal

Jaguar reckon 45 minutes to remove the rear cage. Well, it is possible, and I've done it, but when you throw old exhaust systems and seized bolts into the equation then the decimal point moves to the right at an alarming rate. Allow several hours, especially if you're a beginner!

Only the 'E' type exhaust passes beneath the cage; in all other models, it runs through it, making it a pain to remove unless the exhaust is to be replaced at the same time and then you can just cut it away. What ever the model, thereafter the axle comes out in the same way.

Safety first

A Jaguar, and its independent rear suspension unit, are both very heavy. Ensure that the car is chocked and placed securely on substantial axle stands positioned forward of the suspension unit on a firm chassis member. If you can't get hold of a transmission jack, use at the very least a substantial trolley jack placed under the diff. unit with a large piece of wood to spread the load (we don't want to buckle our bottom plate, do we?).

The drop

Release everything securing the cradle to the car: handbrake cable, brake hose, exhaust and prop-shaft flange. Then move on to the radius arms. Two bolts hold a check strap to the body and a centre main bolt secures the arm. Don't be alarmed if this bolt shears – that can be dealt with later. The biggest problem will be removing the Metalastic bush from the mounting platform on the floorpan. Heat will not help as it will just produce an obnoxious choking smoke from the rubber. The best way I have found is to place a cold chisel against the inner metal on the bush and apply a sharp, sudden blow with a hammer. This is much easier if the old bush is cut away and the radius arm pushed down. On most Jaguars the radius arm mount forms an inherent weak spot only too willing to sacrifice itself to rust. Watch this area very carefully.

That leaves the four frame mounts and the chances are that these will be so perished that time will have done the job for you. If not, it's a straightforward nut and bolt job, probably easiest when removing the top bolts from the car. Have a hunt around the house or streets for a helper to steady the axle unit and then – keeping your feet clear – lower to the floor and drag out from beneath. It's possible to do this with the wheels in place, giving instant mobility to the unit, but it means hoisting the car to a considerable height to give the extra clearance needed.

I'll leave you with that big, greasy, rusty lump sitting on your garage floor and will return in the next issue to discuss the strip-down. Oh, if you're bored, you could try a power wash or the application of much engine cleaner, which will help make it a more pleasant job when the spanners come out.

All the work involved in this strip down has been entrusted to Alan Slawson, specialist in rear end rebuilds (0277 624295) to whom our thanks goes for help with this feature.

IRS! PART 2

1. Lock the half shaft in position by inserting a screwdriver or similar through the inner universal joint so that the shaft can't spin. Remove the split pin from the shaft end nut and remove the nut. Using a three-eared puller (or the Jaguar type), free the shaft from the hub.

2. Undo a securing nut to the outer lower fulcrum shaft and tap out the shaft using a drift. The hub carrier can now be lifted away.

The Strip Down

Jim Patten continues our step-by-step coverage of rebuilding Jaguar's original independent rear suspension. Here the rear frame is dismantled into component parts.

Okay, okay, you can stop twiddling your thumbs, we're back. Hopefully you should have the rear axle on the garage floor by now, with all the loose dirt and grime removed. With a bit of luck, all those injuries incurred in the axle removal have healed and you're ready for the next step.

The only special tool needed here is a hub-puller. If you're a member of the Jaguar Enthusiasts' Club you will be able to obtain the correct puller from them. If not, then a good quality, substantial, three-legged puller will work just as well. Right, let's get to it.

Read this!
Every care is taken to observe safety rules in these articles, but readers undertake this work at their own risk.

3 Release the lower nut securing the shock absorbers and tap the shaft through to free the units. Remove the upper nuts and bolts and lift the shock absorbers and coil springs away. Do not make any attempt to separate dampers from springs at this stage.

4 Remove the four nuts securing the half shafts to the output flanges and lift the half shaft away.

5 This shows the shims between the flanges used to obtain the correct wheel camber.

6 This shows the shim at the outer end of the half shaft used to achieve the correct end float for the hub bearing. These will be discussed further during the assembly procedure.

7 Remove the nuts holding the lower wishbone to the frame and tap out the shaft.

8 The shaft can now be lifted away. Don't worry about the washers and seals that fall; these will be replaced anyway later on.

9 Remove all the brake pipes and the handbrake linkage spring.

10 Cut the locking wire on the bolts securing the differential case to the frame and remove the four bolts.

11 Once the nuts and bolts holding the bottom plate have been removed, the frame (or cage) can be lifted from the differential unit.

12 Now the brake area is visible. See how this handbrake pad has slipped from its mount.

14 Release the two pivot bolts securing the handbrake caliper and remove.

15 Pass a socket through an access hole in the disc and remove the two caliper securing bolts. Remove, noting that there are shims between the caliper and the bracket. New ones will be used during re-assembly to centralise the caliper.

16 Tap the disc off using a soft-faced hammer.

13 Undo and remove the handbrake caliper retractor plate. The right-hand arm shown has fractured through fatigue. Both will be replaced as a matter of course.

This is the stage where those preparing cars to show standard may like to take the opportunity to arrange for the frame to be treated (possibly powder coating) and other minor items sent away for plating. Do arrange this in advance as you might find yourself delayed by the plater.

It ain't 'arf easy taking things apart. Let's rendezvous in the next issue and take a look at what we have; then we can get a little more serious. In the meantime, try to keep everything labelled and 'jared' to make life a little easier when the reconditioning and re-assembling starts.

All the work involved in this strip down has been entrusted to Alan Slawson, specialist in rear end rebuilds (0277 624295) to whom our thanks go for his help with this feature.

IRS! PART 3

Component parts dismantled

Jim Patten follows the stripdown of Jaguar's pre-XJ40 independent rear suspension, here showing how parts are prepared to receive new bearings and bushes.

"What a pile of junk. You'll never do anything with that lot, it's ready for the scrap heap." If you're not careful you could so easily fall into Mr Nosy Parker neighbour's trap and feel pretty dispirited by now, surrounded bits and pieces. But take heart, dear Jaguar disciple, because by the time we've finished your neighbour will think you mad for fitting such a beautiful assembly back on a car.

Well, we're knee deep in springs, shafts and wishbones. Let's see if we can make some sense of it. Leave the calipers to one side, they can have their turn in a later issue. We might even get into some re-assembly this time round, always a positive step. You will need a set of spring clamps but they are easily available at your local spares shop.

1 Measure the disc to see if wear exceeds manufacturer's limits. Check for deep scores or cracks. There's no way to tell if they are warped or distorted without spinning them on a lathe or similar. If they appear to be good, have them slightly skimmed anyway, just to make sure that they run true. But you may decide to do what we did – although these were fine, we replaced them anyway as the cost is relatively low.

2 With the spring/damper unit firmly secured in a vice, attach the spring clamps and gradually compress the springs tightening each clamp progressively. Note! Road springs are dangerous – always use the correct equipment and if you do not feel competent, leave this job to a professional.

3 The retaining blocks can now be removed. They are very similar to valve spring retaining cotters in a cylinder head. The clamps can now be slackened and the spring removed. Unless the dampers are known to be as new, discard and replace with new. Don't just pump them up and down and pass them as fit if they feel tight, that isn't an accurate test.

4 One of our spring seating rings had corroded beyond use, so Alan turned another in the lathe. However, they are available at one of the many specialist Jaguar spares outlets.

5 The hub will still have an inner bearing race attached and will need to be removed. Secure the shaft end in a soft jawed vice and drive the race from the shaft with a drift. This can be a very difficult operation and if you get stuck you may have to use the facilities of a local machine shop.

6 Lay the hub carrier on the bench and tap out the outer bearing seats using a suitable drift. If the seats just fall out then the bearing would have been spinning in the carrier and you will need to source another carrier. Luckily it is fairly common across the range but do check that it is the same carrier. If not, change in pairs.

7 Tap out the lower fulcrum pin bearing seats using a longish drift. Slots are machined in the carrier to give access to the seat.

Read this!
Every care is taken to observe safety rules in these articles, but readers undertake this work at their own risk.

8 It's actually easier to clean and paint the lower wishbone with the old bearings still in place. This way, no stray paint can enter the bearing seats. There are two bearings in each fork so that's four per wishbone. To push out the old bearings choose a socket that is larger than the diameter of the bearing. Then select another of exactly the same diameter as the bearing. Place the sockets either side of the wishbone fork and position in a vice so that when it is tightened, the smaller socket will push both bearings out and into the larger socket. What a lot of words for such a small job.

9 Remove the grease nipple on the inner, lower wishbone and make sure that the hole is clear. Old grease will have congealed at the base and solidified to defy new grease entry. Discard the grease nipple as this will later be replaced with new.

10 Place a new bearing into place and press home in the vice. Repeat this operation for the other side. Early cars had a wider fork end and used a spacer between the bearings.

11 Before doing anything with the drive shafts, use a number punch or other means of identification and mark each yolk with its mate. Remove all circlips. Most will need a jar with a hammer and drift to loosen them from their grooves.

12 Deliver a series of blows with a soft faced hammer to drive the yolk down on the universal joint.

13 The cup can now be lifted away.

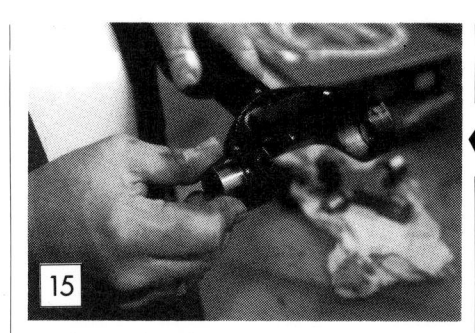

14 Manoeuvre the yolk away from the universal joint and repeat operation 12 for the other half of the universal joint.

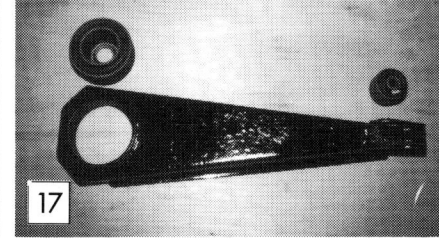

15 Remove the cups from the new universal joints and place one in position and tap home (see how good our newly painted shaft looks). Check inside the cup to see that none of the needle rollers have fallen out of place. Modern kits are pre-greased and don't need any added. Earlier types fitted with grease nipples can be greased after fitting.

16 Place the universal joint in position and slide into the fitted cup. Place another cup into the vacant hole. Push the two cups home using a vice. You will need to use a spacer (small socket or even a nut will do) to push the cups past the circlip groove. Check that the joint moves freely and then fit new circlips. Do not be surprised if a needle roller falls and obstructs the joint. If this happens then the cups have to be extracted, the roller put back in place and the cups pushed home again. Fit the other section of the driveshaft to the universal joint ensuring that all pre-punched numbers correspond.

17 The radius arm bushes are an extremely tight press fit. The old rubber will most likely have fallen out and the best way to remove the metal section is to use a fine, sharp cold chisel and collapse the outer metal part of the bush inwards. Unless that is, you have access to a press, in which case simply press out the old bushes.

18 Alan has devised a simple tool for installing the bushes. The bottom section shown is the same diameter as the eye in the radius arm. The top is the same diameter as the bush. Through the two runs a threaded rod. The bush is pulled down by the top part of the tool when the thread is tightened. Any machine shop will press the bushes in for you for a matter of a few pounds.

19 Both ends of the radius arm are treated in the same way but make sure that the hollow section of the large bush is to the front of the arm.

That little lot is a nice leap forward, you should feel like you're getting somewhere. In the next instalment we start to look at setting things up, end float, pre-loads, all that sort of thing. So I suggest that in the meantime you prepare yourself for the deep thinking to come by taking long walks breathing good fresh air or by eating fish or whatever you do to stimulate the brain cells.

All the work involved in this strip down has been entrusted to Alan Slawson, specialist in rear end rebuilds (0277 624295) to whom our thanks go for his help with this feature.

Kelsey Classic Car Storage

Providing some of the finest facilities in the UK, Kelsey Classic Car Storage is located in a quiet setting, only 15 miles from central London and just 10 minutes from the M25 motorway network. Access to the major international airports could not be easier, and the busy Biggin Hill airport is just five minutes away.

Safeguarding the cars in our care has been a major consideration. With the owners living on-site and the use of the latest high tech security systems, 24-hour cover is provided, making this the very best environment for the storage, protection and maintenance of classic and cherished cars.

Features include a rolling road, full de-humidifying system and private exercise road.

In addition to basic car storage from £23.75 (+VAT) per week, a wide range of other services are available, from simple valeting to a full car care and maintenance programme, MoT testing and a delivery and collection service for both cars and drivers.

Administered by classic car owners with a sympathy for the care and well-being of fine cars, it is truly a facility run by enthusiasts for enthusiasts.

www.kelsey.co.uk/storage or for a brochure with further details contact:

Kelsey Classic Car Storage

Tel: **01959 541444**.   Fax: **01959 541400** (International calls: 0044 1959 541444)

or e-mail: **storage@kelsey.co.uk**

Jaguar World, September/October 1993

IRS! PART 4

End floats & pre-loads

Lower Fulcrum Shaft Bearings and Adjustment

Jim Patten follows the strip-down of Jaguar's pre-XJ40 independent rear suspension. This instalment covers the fitting of bearings and adjustments.

"That boy there, at the back, Simkins major, pay attention. If you thought Latin was bad then you are in for a very nasty surprise – it's the IRS period now!"

That's right, it's brain teaser time but, if you took my advice and made your preparations, then your mind should be ready for anything. Actually it's quite easy really; what we are setting out to achieve is a hub bearing end float of 0.002in-0.006in and a fulcrum shaft bearing pre-load of 0.000in-0.002in. The hub bearing adjustment is possibly one of the less understood areas on this rebuild. So often I have heard of people trying to effect adjustment on the nut by the technique of 'wind up as far as it goes and back off a turn'. Wrong, and if you've done that, into the corner you go with the pointed hat on.

The correct method is to use shims behind the bearing as will be outlined in the following sequences. We have adopted a slightly different method from the manual but the end result is the same. But first the new bearings have to be pushed in. If you have not the facilities of a press then a local machine shop will do the job for you for a nominal sum.

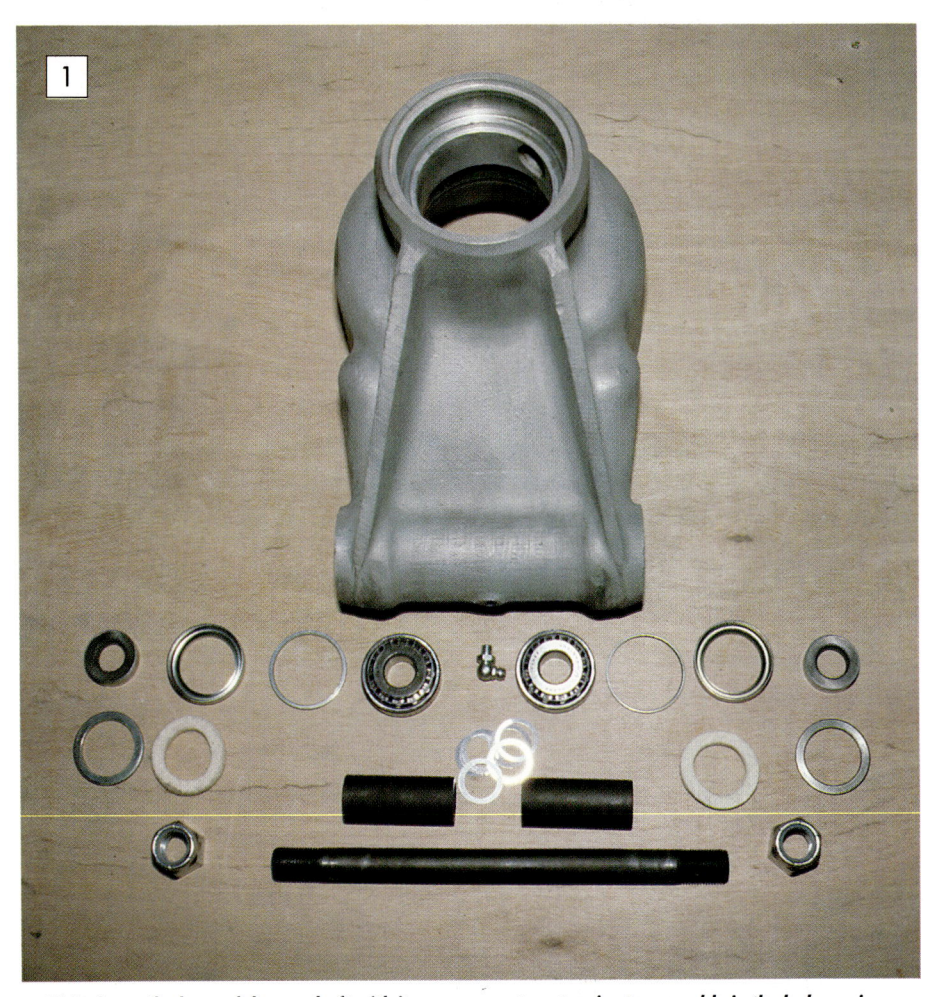

1 *This shows the lower fulcrum shaft with its component parts prior to assembly in the hub carrier. As there is a bearing in each end of the shaft, each side is a mirror of the other. They are fitted on the shaft as follows. Left to right are: the shims to centralise hub carrier in the wishbone, retaining washer on seating ring for the oil seal, container for oil seal and felt seal, seating ring for oil seal, bearing, spacer sleeve and shims fitted between sleeves for pre-load.*

2 *To set up for the assembly procedure it will be necessary to knock-up a simple jig. Use a piece of plate steel approx. 7in x 4in x 3/8in. Drill and tap a hole to receive the lower fulcrum shaft so that the shaft can be retained as shown.*

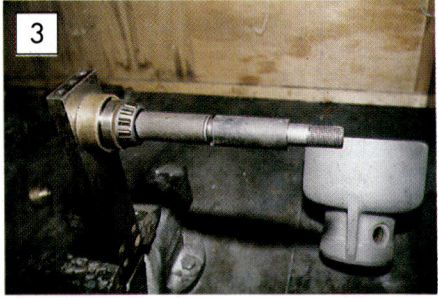

3 *Position a spacer to an approximate size of the lower wishbone outer forks on the shaft first followed by the seating ring for the oil seal, bearing, spacer sleeve and an excess of known size shims and the second spacer sleeve.*

4 Press in the two end bearing tracks and then slide the hub carrier over the shaft and push the other bearing and seating ring on. Fit a large washer (the inner wishbone fork outer thrust washer is perfect) and the nut and tighten to 55lb/ft. Swivel the hub carrier around to settle the bearings and take a measurement using feeler gauges between the washer and the machined carrier face.

5 We are looking for a mean pre-load of 0.001in. We obtained a reading of 0.013in (end float). To get the required pre-load, shims to the value of the end float and the pre-load figure, that is 0.013 + 0.001 = 0.014 should be removed from the centre of the fulcrum shaft.

6 Use a dummy shaft (that is one that measures the same as the width of the hub carrier) to assemble the components.

7 Build up the shaft from the centre out: correct shims, spacer sleeve, bearing, seating ring for felt seal, spacer between bearing and container, felt seal, container for felt seal and tap the retaining washer into position.

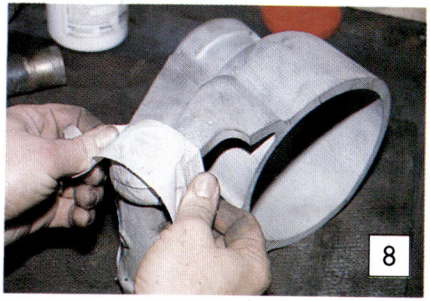

8 Bind the ends using a good quality tape.

Hub Bearings and Adjustment

9 Position the oil seal track before pressing the inner bearing on to the hub using an hydraulic press.

10 Tap the outer race tracks into position in the bearing housing.

Offer the hub (with the bearing greased) to the hub carrier.

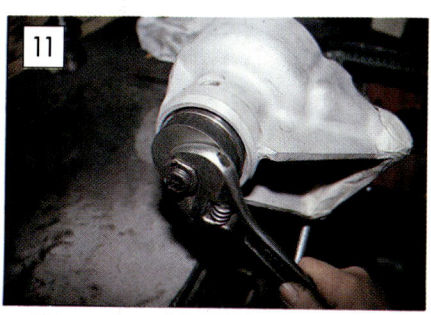

11 Using a piece of threaded rod and two thick spreading washers of the same diameter as the bearing seats, place through the housing with the washers placed each side of the seats and wind until the seats are fully pressed into place. The pressure needed here is considerable – if you can use an hydraulic press then do so. Add outer seal.

13

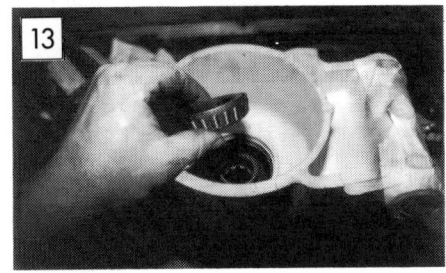

Place the other bearing (greased) in the rear of the hub.

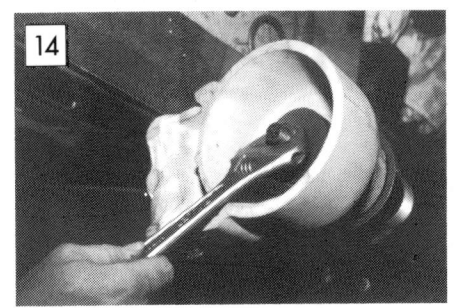

14

Back to our piece of threaded rod. This time it is needed to pull the bearing home.

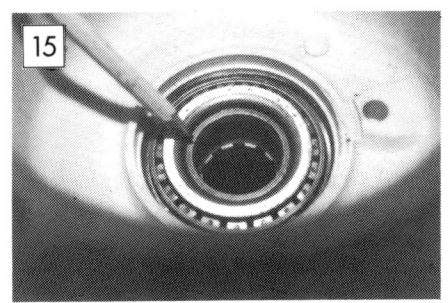

15

With both bearings home it will be seen that the hub shaft sits lower than the bearing centre. We need to establish the thickness of shim required to sit on the hub shaft and be proud of the bearing centre by 0.002in – 0.006in to give the required end-float.

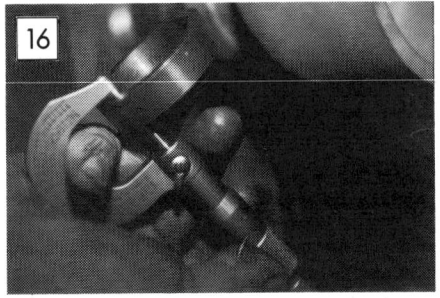

16

Take a piece of machined brass to sit on the bearing centre. Accurately measure its thickness using a micrometer.

That concludes this gruelling exercise. It might seem a little daunting but honestly, as soon as you get the idea it is almost (but not quite) simplicity itself. This episode's advice is to toddle off to the off-licence for your favourite tipple (I still had a bottle of Cabinet Sauvignon from the Guenoc winery) and forget about the IRS until the next issue when we look at the brakes.

All the work involved in this stripdown has been entrusted to Alan Slawson, specialist in rear end rebuilds (0277 624295) to whom our thanks go for his help in this feature.

17

Read this through a few times until you fully understand what is happening.

Using a depth micrometer take a measurement (a) from the top of the brass ring to the base of the hub shaft and note it (it being 0.622in here) down.
Subtract the thickness (b) of the brass ring (ours was 0.5in) from 'a' to give the gap (c) (0.122in in our case) between hub shaft and bearing centre.
Add to this the required end float (d) to give the thickness of the spacer required (e).
Spacers are available in 0.003in increments and are lettered for identification. They start at A for 0.109in and finish at R 0.151in.

Now putting our algebraic equation together we have:
$a - b = c$.
That is: $0.622 - 0.500 = 0.122$.
$c + d = e$.
Now we have to work slightly backwards in our sums to choose a spacer that will give us the end float.
Looking at the list we find that we can use F (0.124) or G (0.127).
We had a G and that gave us an end float of .005in, $0.122 + 0.005 = 0.127$.

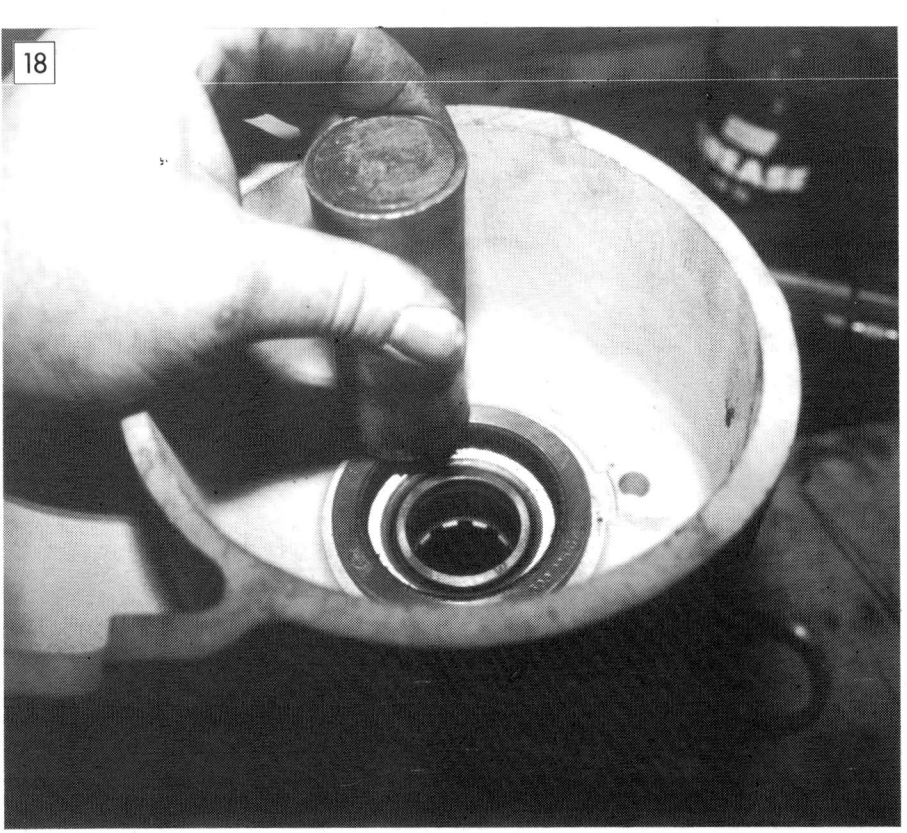

18

With the spacer in position, fit the inner oil seal followed by its seating ring (dished side facing in).

S·N·G Barratt

The Heritage Building
Stourbridge Road
Bridgnorth
Shropshire
WV15 6AP

Tel: 01746 765432

Fax: 01746 761144

RING OR WRITE FOR AN ITEMISED QUOTATION OR PRICE LIST TODAY!

S N G Barratt is a long established company with its roots firmly established with the enthusiasts. We have been retooling, buying final production runs, investing in obsolete parts and accumulating vehicles for restoration and breaking for over a decade now. Our stock is made up of 10,000's of parts to make your Jaguar stop, go or just look great together with many rare and vital components to keep one of Britains finest marques alive. Make us your first call for all your parts needs and see how helpful we are!

How to order:

By Telephone: call us between 8.30am - 5.30pm Monday to Friday and 8.30am - 12 noon on Saturday. One of our expert sales team will be happy to take your order and assist with any queries you may have.

By Post: Send us a list of parts your require along with your remittance or credit card details.

Shipping and Cost: To ensure that your order is delivered quickly and safely, our carrier will take up to 30kg on a next day basis for a flat rate. If security and speed are not an issue, we can send via the postal service.

Western Classics
JAGUAR SPECIALISTS

SERVICING • ENGINE REBUILDS • TUNING • RESTORATION

MAJOR/MINOR SERVICING ★ COMPLETE/PARTIAL ENGINE REBUILDS ★ MAJOR/MINOR REPAIRS ★ GEARBOX/AXLE REBUILDS
★ MAJOR OVERHAULS ★ SUSPENSION REBUILDS

We rebuild all types of Jaguar engines from 1935 to 1995 from 4 cylinder to 12 cylinder. All engines can be built to your requirements either to original specification or tuned. All engines run on test bed on completion.
Also bodywork, painting and trimming carried out to your specification.
We have many years experience with all types of JAGUARS.
WE ARE THE EXPERTS.
Please phone Alan or Fred for free advice/estimates. We look forward to hearing from you.

7/8 St. GEORGES WORKS, SILVER STREET,
TROWBRIDGE, WILTS BA14 8AA UK.
Tel: 01225 751044 Fax: 01225 751641

Engineering, workshops, manufacturing and parts suppliers - UK EUROPE, WORLDWIDE

XK, 'E' TYPE & MK2 SPECIALISTS

We are one of the UK's larger suppliers of parts for all JAGUAR models from 1948 onwards. We manufacture many parts ourselves with the objectives of providing both good quality and excellent availability.
BRITISH MOTOR HERITAGE approval means we can always be relied upon to give a professional and complete service.

We would be pleased to carry out servicing and other major work to your car.

BRITISH MOTOR HERITAGE APPROVED

We can also supply parts for AUSTIN HEALEY & RANGE ROVER

CATALOGUE AND PRICE LISTS AVAILABLE

SC Parts Group Ltd
13 Cobham Way Gatwick Road Crawley W. Sussex RH10 2RX UK
Telephone 0293 547841/4 Fax 0293 546570

67

IRS!
PART 5

Jim Patten follows the strip-down of the pre-XJ40 independent rear suspension. In this issue, the rear calipers are rebuilt.

Stoppers

1

Remove the pad cover/stop plate, retaining pin and brake pads and, left in the space vacated, undo the bolts securing the pad supports. There are four semi-circle pieces, one top and bottom of the pads.

2

With the caliper secured firmly in a vice, remove the detachable cylinder from one side of the unit. The other cylinder is in the housing itself. Ease the pistons out of the bores using compressed air (be careful) or a pair of thin end levers, try not to slip as you could make a nasty mess of your hands. Inside the cylinders will be the sealing ring; this can be flipped out using a small screwdriver.

Rest period over, it's back to the workbench. This issue is more fiddly than brain teasing but extremely important. Although not playing a role in the IRS workings, the brakes form an integral part of the overall configuration. As far as safety goes, the importance of their condition cannot be overemphasised.

We had a stroke of luck in finding some new front calipers but the rears will have to be reconditioned. As a seal is recessed into the bore, most wear will be on the piston and little or no attention will be needed on the cylinder apart from a thorough clean. Our pistons were in pretty bad shape, so we had new ones made from stainless steel by Tim Smith of Avon Engineering (Tel: 0279 816541).

The handbrake mechanism on this 'S' type suspension unit is of the self-adjusting variety. Earlier cars ('E' type and Mk X) were adjusted manually. With the auto type, the handbrake operation pulls the operating lever (and pads) on to the disc. If there has been any wear in the pads or disc, the pawl rotates the ratchet adjuster nut and draws the bolt inwards, compensating for the wear. Clever job. However, this assembly sits high and isolated in the centre of the rear axle and, neglected, it often fails to operate. Then the myth develops – Jaguar can't build a car with a good handbrake. Oh, yes they can! Read on and yours will work properly too.

Those bright and eagle-eyed will have noticed that the job is not quite finished. That's true, but it's as far as we can go in this issue without bolting the disc to the axle and we're not ready for that yet. When we have checked out the differential unit,

Warning! Braking systems affect safety. If you are not confident in your ability to do the job correctly, take your car to an expert. This article is intended for the guidance of competent mechanics.

3 With the cylinder thoroughly cleaned the assembly can begin. Coat the inside of the cylinder and the new rubber seal with brake fluid. Ease the seal into its recess in the cylinder wall, making sure that it does not twist and is firmly seated.

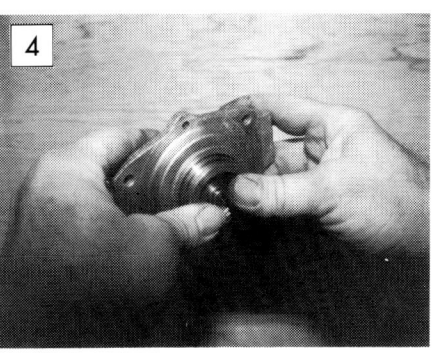

4 Using our new caliper piston (or the old one if it is in a new condition), coat with brake fluid and ease into the bore. Be very careful not to nick the rubber seal. The brake fluid should help it slip into position. It may well be a tight fit as it goes down.

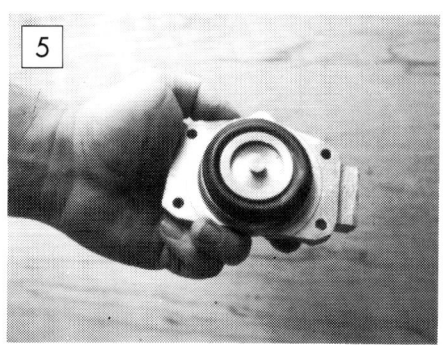

5 Fit the dust cover over the grooves in the body and the piston. Repeat all operations for the other bore. Access will be a little more difficult due to the nature of the caliper body but it should not present any problems.

6 Bolt the caliper cylinder on to the main caliper assembly. We used new bolts and washers; the cost is minimal. Use firm and even pressure but don't overdo it – it's a nasty job extracting broken bolts.

7 To gain access to the handbrake adjuster mechanism, remove the cover by releasing a central screw and a top pin. This will also free the operating lever from the pad carrier.

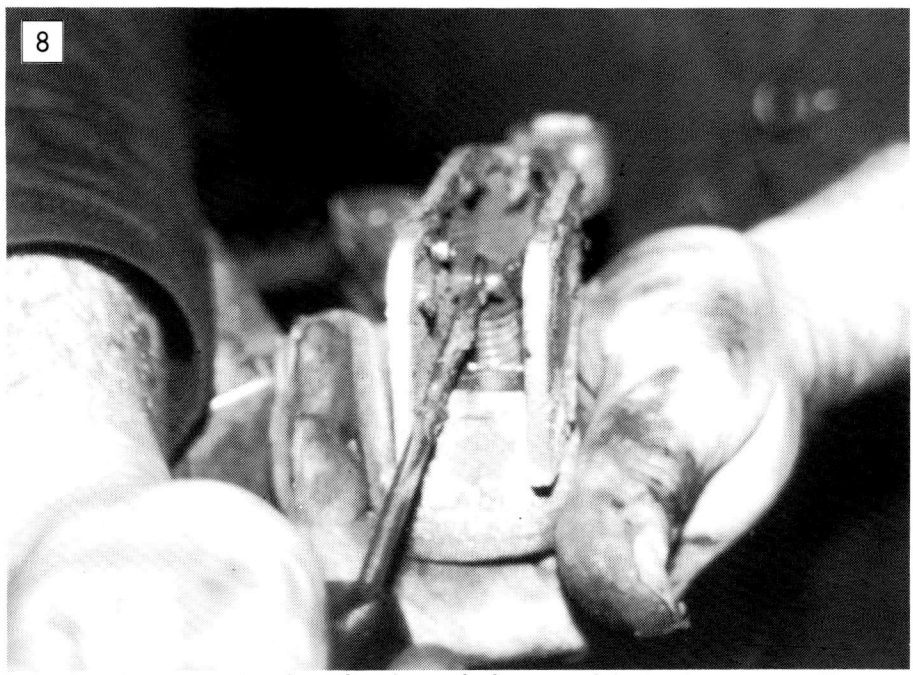

8 Prise off the return spring. The anchor pin can also be removed, freeing the cover assembly.

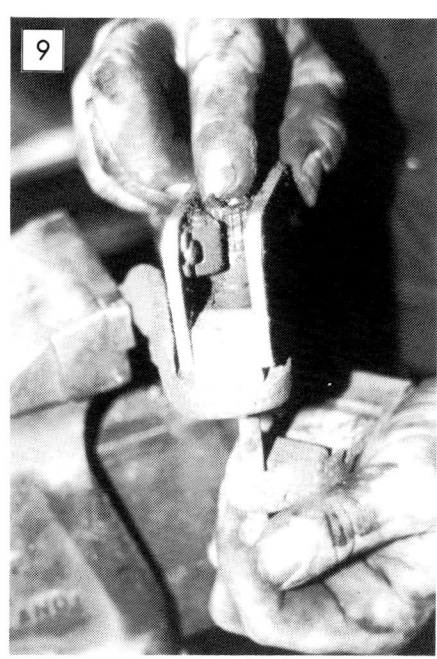

9 Remove the securing split pin and then the bolt through the two pad carriers and operating lever. Ours had become badly bent and would be replaced. The pawl and adjusting nut can be seen clearly here.

Jaguar World, November/December

10

Remove the pawl tension spring, adjusting nut and pawl assembly.

12

11

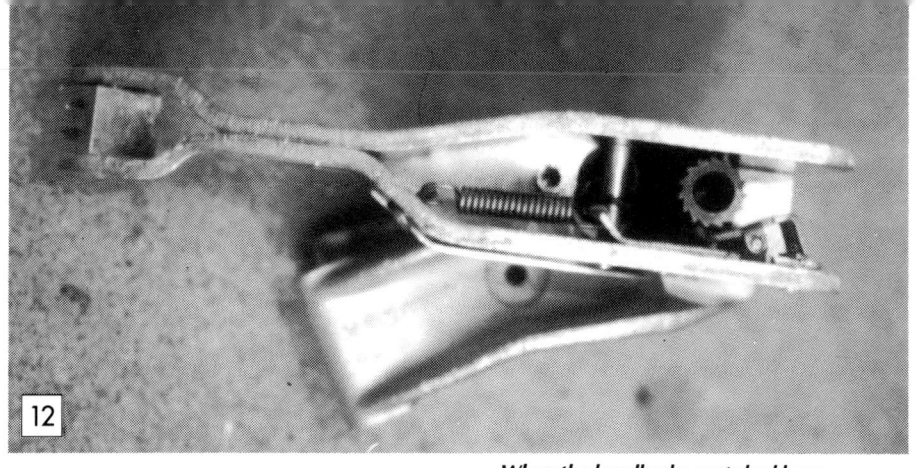

When the handbrake parts had been cleaned (we had them plated), assembly could begin. Replace the friction spring, adjusting nut and pawl assembly. The anchor pin and tension spring could now be put in place.

The handbrake pads are secured by a bonded plate that slides over a pin on the pad carrier. This pin passes through the pad carrier and is secured by a nut. The pads can be pulled off in the normal way and then the pin released for examination. If in doubt replace. We did.

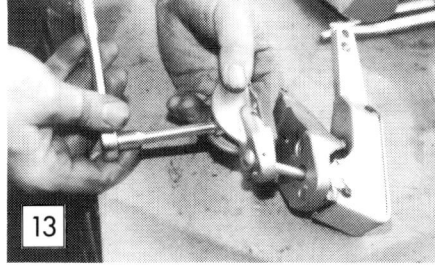

13

With the bolt securing the two pad carriers and operating lever in place, the pad retaining pin can be positioned.

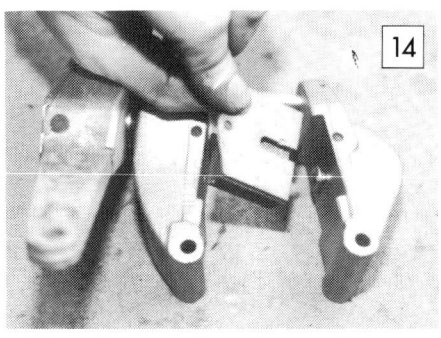

14

The pads can now be pushed over the retaining pins and the nuts tightened.

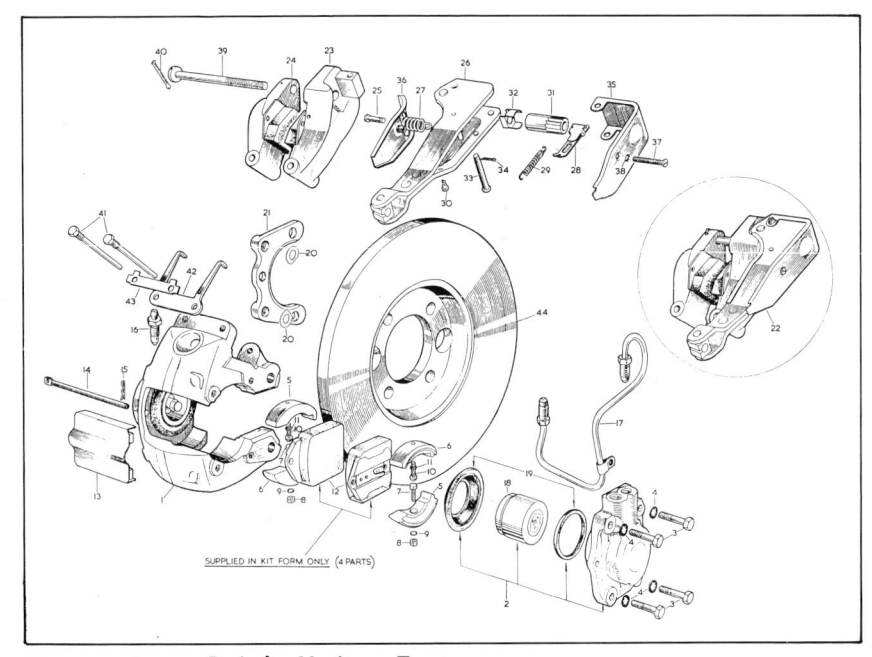

Courtesy Jaguar Daimler Heritage Trust.

1. Caliper and piston assembly. **2.** Piston and cylinder assembly. **3.** Bolt securing cylinder to caliper. **4.** Washer. **5.** Pad support RH. **6.** Pad support LH. **7.** Bolt securing pad supports at bottom of calipers. **8.** Nut. **9.** Washer. **10.** Screw securing pad supports at top of calipers. **11.** Washer. **12.** Friction pads. **13.** Stop plate assembly. **14.** Pin retaining pads and stop plate. **15.** Clip retaining pin. **16.** Bleed screw. **17.** Bridge pipe. **18.** Piston. **19.** Seal kit. **20.** Shims. **21.** Adaptor plate. **22.** RH handbrake mechanism. **23.** RH inner pad carrier. **24.** RH outer pad carrier. **25.** Anchor pin. **26.** Operating lever. **27.** Return spring. **28.** Pawl assembly. **29.** Tension spring. **30.** Anchor pin. **31.** Adjusting nut. **32.** Friction spring on adjusting nut. **33.** Hinge pin. **34/** Split pin through hinge pin. **35.** Protection cover rear. **36.** Protection cover front. **37.** Bolt securing covers. **38.** Washer. **39.** Bolt securing pad carriers. **40.** Split pin retaining bolt. **41.** Bolt securing handbrake mechanism. **42.** Retraction plate. **43.** Tab washer. **44.** Disc.

Our thanks go to Tim Smith of Avon Engineering (Tel: 0279 816541) for such a fine job of remanufacturing the caliper pistons in stainless steel. All the work in this series has been entrusted to Alan Slawson (Tel: 0277 624295), specialist in rear end rebuilds

NEXT ISSUE

We check out the differential unit and hang the brake discs.

WE GET TO THE PARTS OTHER DEALERS CANNOT REACH

When it comes to caring for your Jaguar or Daimler an official dealer is beyond compare. The same can be said of Genuine Jaguar Parts. Naturally, we offer you the widest choice backed by direct computer access to Jaguar's vast centralised stock. So even if you want rarer parts like an E-Type hed light or a Series III propshaft, we can get it for you quicker than anyone. And with our competitive prices, all of this is within easy reach.

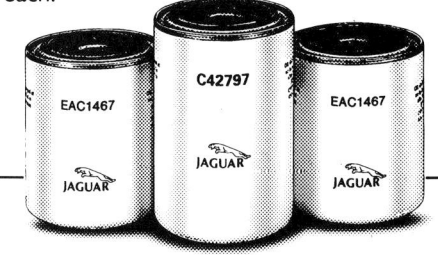

HARVEY HUDSON of WOODFORD
JAGUAR PARTS SPECIALIST

Mail Order - Fast, reliable service.
Daily deliveries in London and Essex

PHONE: 0181-6644. FAX: 0181-989 6025.

LARGE STOCKS OF GENUINE JAGUAR PARTS XJ40 - XJS - SERIES I, II & III.

Special terms for members of Jaguar Drivers Club and J.E. Club

JAGUAR Daimler OFFICIAL DEALER

Harvey Hudson & Co. Ltd., 50 - 56 High Road, Woodford, London E18 1AS.

JAGUAR GENUINE PARTS

GUY SALMON JAGUAR

PARTS DIVISION

**Kingston House Estate
Portsmouth Road
Surbiton Surrey
KT5 5QG**

THE JAGUAR SPECIALIST

**Telephone:
0181-398 7646
Facsimile:
0181-398 9064**

ONE NAME THAT GUARANTEES EXCELLENCE

IRS! PART 6

Jim Patten continues the rebuild of Jaguar's independent rear suspension

If I mention the differential it's only because this great heavy chunk of workings ought to be checked before it takes its place back in the rear suspension. If there is a problem then let's know about it now before it's too late. While a differential rebuild is beyond the scope of this feature, possible faults to spot include oil leaks from the differential seals (they can get cooked by those inboard brakes), worn differential bearings and burnt clutches in the Powr-Lok device betrayed by brown tainted oil. Also remove the back cover and check the crown wheel and pinion for backlash. Comforted by thumbs up or with a rebuilt unit, you can continue the reassembly with confidence.

Differentials and discs

We opted to fit new brake discs although our existing parts were in superb condition. If we'd decided to use them, we would have had a machine shop give them a precautionary light skim. But so reasonably are the new items priced that the extra expense over the resurfacing cost seemed worthwhile.

The rear wheel camber is determined by the use of shims between the driveshaft and output shaft. As this adjustment is carried out with the axle in the car with a given laden weight, we assembled with a token known number of shims, just as a datum point.

1 With backplate removed from differential casing, a dial gauge indicator is mounted with the pointer against the crown wheel face. Spin the wheel by hand. Run-out should not exceed 0.005in. The backlash figure will be etched on the drive gear. Position the gauge with the pointer against a drive gear tooth in line with the direction of travel. Move the wheel and check that the backlash does not exceed 0.004in. of the etched figure. Visually check all teeth for signs of odd wear or chipped teeth. If all is well, bolt back cover using a new gasket.

2 New bearings and 'O' ring have already been pushed on to this output shaft.

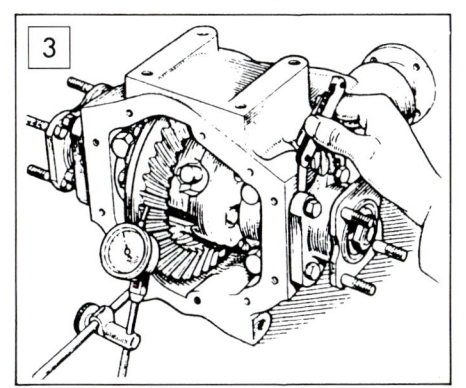

3 Fit the shaft to the diff. housing without shims and measure between the shaft and housing. Fit shims to achieve an end float of between 0.001-0.003in and fully tighten the shaft to the housing.

4 The caliper mounting bracket is fitted to the rear of the output shaft.

5 Using temporary spacers, fit the disc in place and fully tighten the nuts. If using an old disc, check for run-out with a dial gauge as illustrated.

With the differential in an upright position, lower the cage over the top and secure using the special tapered bolts. When they are fully tightened, join each bolt with locking wire.

6 The caliper can now be fitted to its carrier and centralised over the disc, using shims to adjust.

7 Fit the handbrake caliper on to the main caliper and secure with the long pivot bolts and lock tabs. The brass spring yoke goes under the bolt heads with the prongs located in the handbrake calipers.

9 Bolt the lower mounting for the inner fulcrum shaft to the differential casing. There are shims used. Here it is a case of putting back what was taken out as the camber adjustment will be made on the drive-shaft. Secure the bolts with locking wire.

10

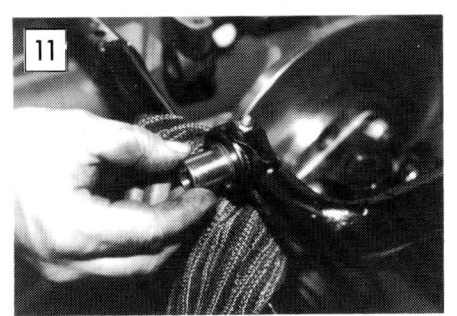

11 Take a bottom wishbone and grease the bearings (inserted in a previous episode) and insert the bearing tube.

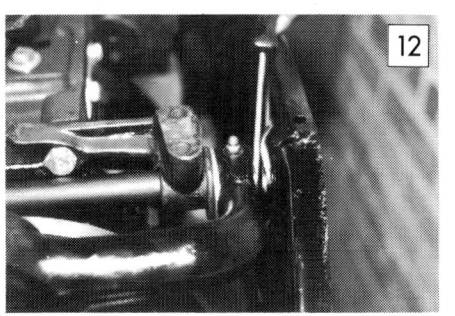

12 Position the lower wishbone distance tube between the two bracket ends on the mounting. Place the lower wishbone on the mounting and juggle the thrust washer (inner), sealing ring, retainer and thrust washer (outer) each side of the wishbone forks (four sets).

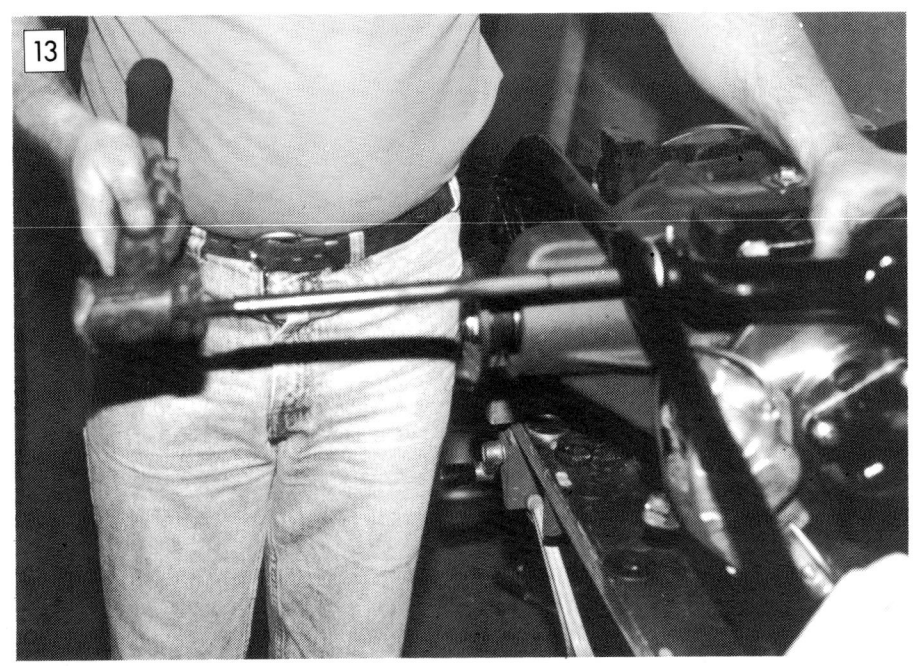

13 Coat the inner fulcrum shaft with Copperslip or similar and tap home through the lower wishbone, mounting and distance piece. There may be a fair amount of jostling involved as everything lines up. Replace the two end Nyloc nuts.

14 Coat the radius arm retaining bolt with Copperslip and fit the radius arm to the lower wishbone. Make sure that the correct bolt is used as they vary. Later ones will not fit early cars and vice versa.

Remove the temporary spacers securing the brake disc and insert a couple of shims. Note the thickness as this will be taken into account when the camber adjustment is made. Offer up the drive-shaft and bolt it securely into position.

15 Place the new shock absorbers in a vice and compress using quality spring compressors. Take extreme care. If these should give way or slip the consequences would be disastrous. Insert the seat and retainers. When they are in position, slacken the clamps making sure that the seat is firmly in place before removing the clamps. If you do not feel competent to undertake this work, get an expert to do it for you.

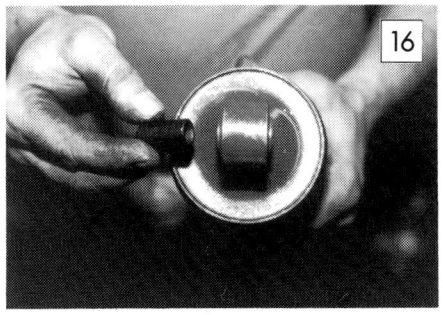

16 Don't forget to fit the bush on the eye of the shock absorber.

17 Fit the shock absorbers first into the cage and then to the lower wishbone. Bolts secure the top and a single shaft the bottom. The usual error here is to work on the cage upside-down and then fit the shock absorbers the right way up. When the cage goes in the car, the shock absorbers are the wrong way up. Don't feel too bad – I did it that way myself first time.

What a pity, we've run out of space to go any further in this issue. Be patient, we should just about finish the job in the next when we fit the hubs and carry out the brake plumbing.

All work in this series has been entrusted to Alan Slawson (Tel: 0277 624295), specialist in rear end rebuilds.

IRS! PART 7

Jim Patten concludes the rebuild of Jaguar's independent rear suspension

The home straight...

We're there at last, the long haul is almost at an end - just the hub carriers and driveshafts to go, with a little bit of plumbing around the brake parts. As an aid, we are reproducing an exploded diagram of the rear suspension along with part numbers and the retail prices as at December 1993.

Although we have concentrated on the 'S' type throughout this series, the parts are fundamentally the same for virtually all IRS Jaguars from the 1961 'E' type through to the XJS and Series 3 XJ saloons. Some models, notably the 'E' type, are fitted with a rear anti-roll bar - wearing parts here are simply a couple of bushes in the bar links and two further roll bar rubbers securing the bar to the rear bulkhead. We have refrained from listing brake parts as well because these vary so much over the different models.

Generally we have been going strictly by the book but during the reassembly of the rear end, you may like to consider an adaption to make the job of bleeding the rear brakes a bit easier. Bailey Brakes (081 459 0307) supply a kit that consists of correct length brake pipe, an adaptor and bleed nipple.

The existing bleed nipple is removed from the caliper and replaced by the length of brake pipe, the other end of which is mounted on the cradle exiting in the wheelarch side. You may have to make a hole in the cradle to secure it firmly or alternatively mount a carrier bracket to the side.

On my last 2+2 'E' type I used a flexible brake hose from the caliper to the inspection plates behind the rear seats where the bleed nipple was fitted. It worked perfectly. Not for purists maybe, but it does save a whole lotta sadness laying underneath the car with brake fluid running down your arm during efforts to "expell the air from the system".

2 Slide hub and carrier over splined end of drive-shaft making sure that the shim, so carefully calculated in a previous issue, is in place. It is important to ensure that the split-pin hole in shaft corresponds to hole in hub. Position hub carrier (with dummy shaft still in place) between the forks of the lower wishbone.

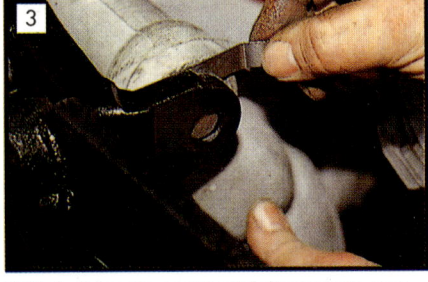

3 With hub carrier pushed fully over to one side, measure the clearance using feeler gauges between oil seal track and wishbone face. Use shims (available in .004") to centralise hub.

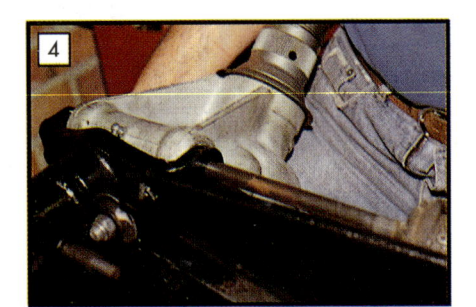

4 Tap lower fulcrum shaft into position, displacing dummy shaft in the process. Replace the Nyloc nuts, tightening to a torque of 55lb/ft.

5 Replace slotted nut and washer to secure hub to shaft. Fully tighten to a torque of 140 lb/ft. Replace split-pin by entering through the access hole in hub. It may be necessary to rock the nut back a little to get the split-pin in.

6 The brake pipes can now be plumbed in. We used the easy to bend, long lasting copper pipes, bought as a vehicle set from Automec (0280 822818). There is a three way junction secured to the frame. One pipe from each caliper and the flexible hose to join the circuit on the body. There is an acute bend to make to one of these junctions and extreme care should be taken to avoid 'closing' the pipe.

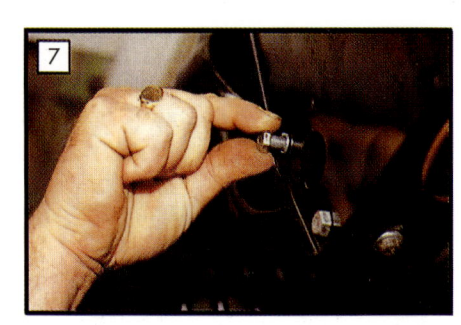

7 Replace the lower pan to cradle by bolting to lower wishbone brackets and around the edge of the cradle. Note that the brake pipe retaining clips running along the front of the frame are secured with these bolts.

8 Replace grease retaining caps on hub carrier. These will need tapping home. Use an old socket to prevent damage to the cap.

9 Covers protecting the four driveshaft universal joints are in half. Position them over the shafts so that the bell end covers the joint and if a grease nipple is fitted, the hole in the cover corresponds. Align the pop-rivet holes and rivet in place.

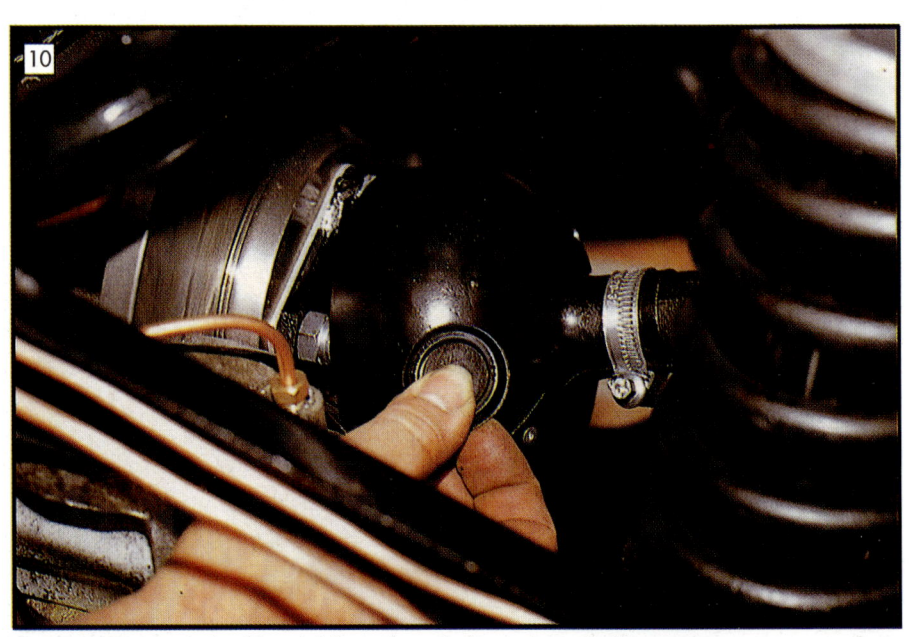

10 Secure cover to shaft with a Jubilee clip and place a bung in the grease nipple access hole.

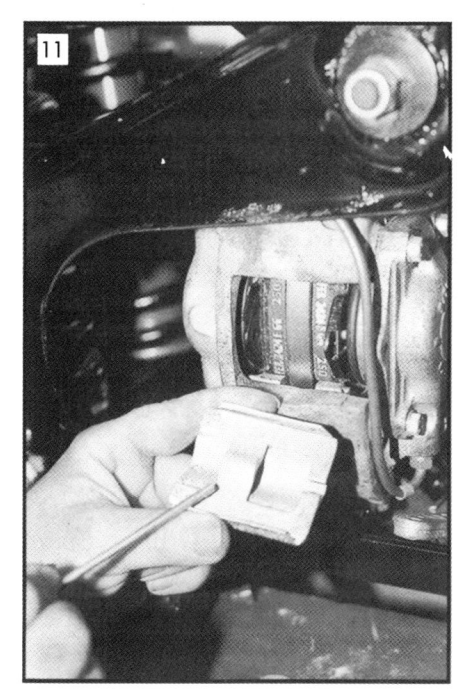

11. Fit new brake pads and in the case of the 'S' type saloon, put the cover in place and secure with the pin. Ours had perished but new ones were obtained from Bailey's Brakes.

12. Here the rear end is positioned ready to be lifted in place.

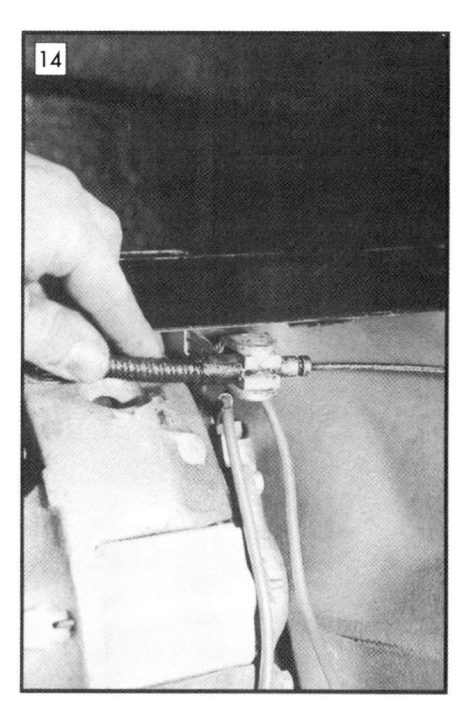

13. It sometimes helps to put a rear mount in place first and then lift the assembly up to it.

14. Before jacking in place, position handbrake cable through eye of first caliper and secure to second with a clevis pin.

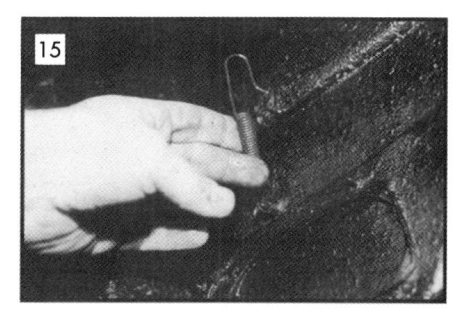

15. Don't forget to replace the spring that supports the handbrake cable. Hook this over the cable only when the rear end is secured.

16. Try to have some help at hand as the complete unit is jacked into place, and take great care as it's heavy! Lay some old tyres around the suspension just in case the worst happens and it slips. Bolt all the mounts up securely and then turn to the radius arm mount. Apply some 'Copperslip' or similar to all bolts and a smidgen on the radius arm mount itself. Don't forget to put the safety bracket over the radius arm bolt and then use locking wire to finally secure the bolt.

There, job done and I hope you feel a better person for it. All that remains is to fit the unit back into the car, an operation that in its basic form, is the same whatever car it is.

If you have the time, clean as much of the body usually hidden by the axle and apply a quality rust preventer such as Waxoyl. To avoid any future problems in removing the cage, coat all bolts and the radius arm locating dome with Copperslip or similar. On the 'S' type/420 saloons, try fitting the first pair of body mounting rubbers to the cradle and the second set on the car. Every little helps when you're trying to jiggle that great lump in place.

Make sure that your jack has a wide head and that you use a thick piece of wood to spread the load across the cradle's bottom pan. Do not choose a piece that exceeds this as it will invade space occupied by the lower wishbone and the unit will become unstable. You'll need some help to keep everything steady as it eases into position. Don't be an idiot and lay underneath wriggling the cradle about to line bolts up, that's a sure fire way to end up in casualty. Try wherever possible to use new nuts and bolts.

On cars fitted with wire wheels, coat the splines with a quality grease. I had an 'E' type some years ago where a previous owner had never bothered, and the only way I could remove a front wheel was to loosen the caliper bolts and free the brake pipe, remove the centre hub nut (this

Jaguar World, March/April 1994

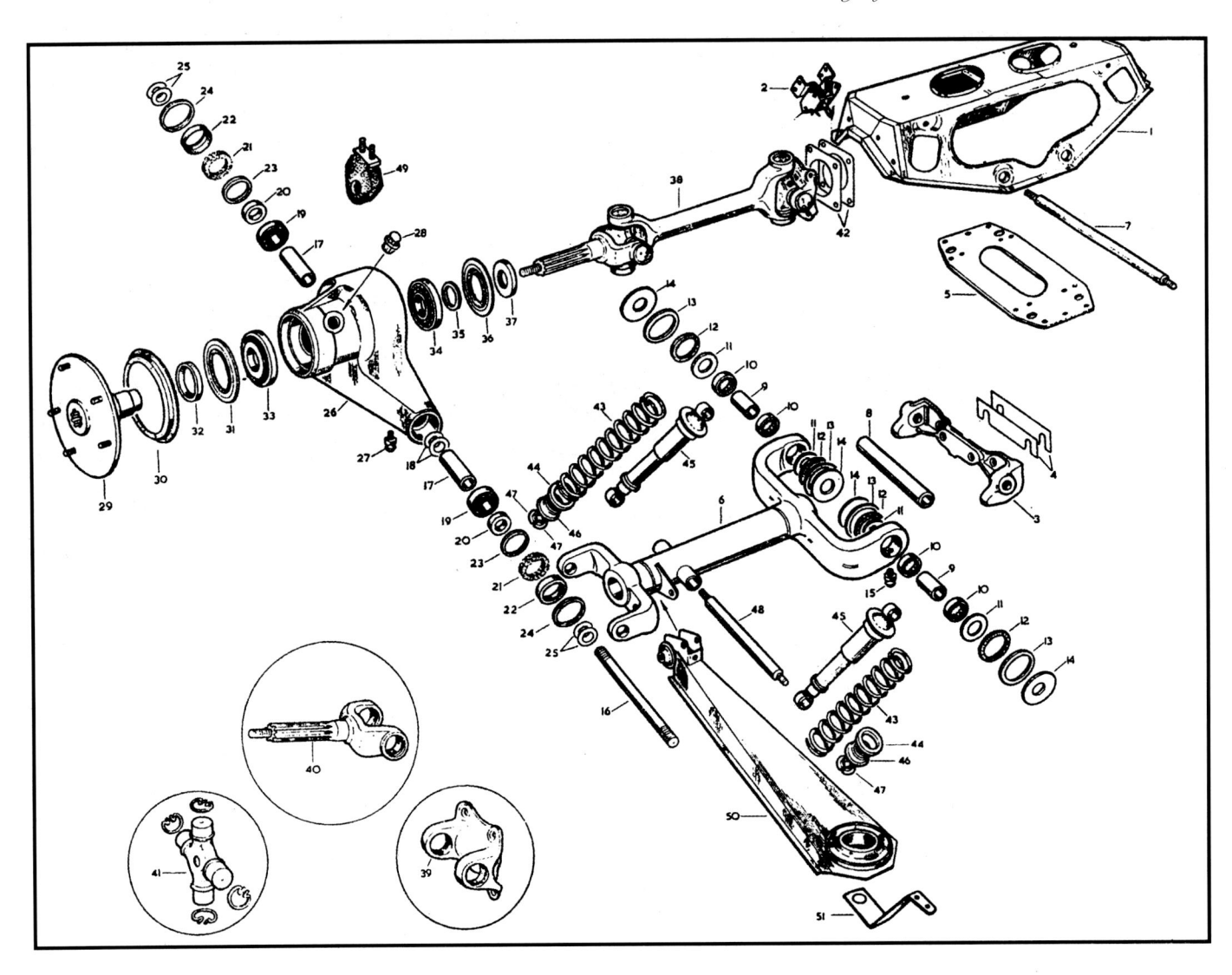

meant hammering a socket over the top and applying a huge amount of pressure to 'shear' the split-pin) and remove the whole thing from the car, physically smashing the hub from the wheel. Everything except the caliper was destroyed for the sake of a bit grease...

Now that you have a fully rebuilt unit, don't neglect it - stick to the service schedules even if that means greasing the joints every six months whether the car is used or not. Keep to that and you will always have a taut Jaguar that will be a joy to use.

In our next overhaul series we will be looking at the refurbishment of the SU HD6 carburettor.

Acknowledgements
Rebuild:
Alan Slawson, specialist in rear end overhauls, 0277 624295.
Brake machining:
Tim Smith, Avon Engineering, 0279 816541.
Shot-blasting:
Mike Carlton-Baker, 0277 821491.
Brake components supplied by:
Bailey Brakes, 081 459 0307.
Brake pipes supplied by:
Automech, 0280 822818.
Other parts supplied by:
Grange Jaguar, Brentwood, 0277 260793
Ken Jenkins (Jaguar Spares) 0909 732219.

Parts and Prices

2	C.17198	Body mounting rubber x 4	£26.25
9	C.17168/1	Fulcrum shaft bearing tube x 4	£9.30
10	C.17167	Needle bearing, wishbone x 8	£4.55
11	C.17166	Thrust washer inner (on bearing tube) x 8	£3.40
12	C.17213	Sealing ring on inner thrust washers x 8	£1.30
13	C.17936	Retainer for sealing rings x 8	£3.30
14	C.17165	Thrust washer outer (on fulcrum shaft) x 8	£3.15
15	C.3044/1	Grease nipple for inner fulcrum shaft x 4	
18	C.16626, /1,/2,/3	End float shims x as required	£0.25
19	C.16029	Bearing on fulcrum shafts x 4	£9.50
21	C.20178	Felt oil seals on fulcrum shafts x 4	£0.79
22	C.20179	Container for oil seal x 4	£4.60
	C.8667/7	Self-locking nut on fulcrum shaft x 4	£0.34
27	C.3044/1	Grease nipple in hub carrier (wire wheels) x 2	
28	C.18124	Grease retaining cap on hub carrier x 2	£2.60
	JLM.9732	Hub bearing kit x 2	£29.00
35	C.19110	Spacer shims on half shaft x as required	
41	10420	Universal joints x 4	£14.00
42	C.16621	Shims between drive shaft and output shaft x as required	£0.25
	C.15349	Nut self locking drive shaft to output shaft x 8	
	L.105/13U	Split pin on hub bearing nut x 2	
	C.23782	Radius arm bush front x 2	£12.20
	C.17146	Radius arm bush rear x 2	£9.50

Total excluding small items (grease nipples, shims, nuts and bolts), brakes, shock absorbers and VAT. £491.22

The above prices are per item and as supplied through a Jaguar main dealer. If you source your parts outside of this sphere, make sure that they are either genuine parts or from a thoroughly reputable manufacturer. We have heard some real horror tales concerning spurious parts. They are not worth the pound or two saved. Shock absorbers and brake parts vary between models and these should be priced as they are needed. Some dealers offer a discount to members of a Jaguar/Daimler car club.

Diagram courtesy Jaguar Daimler Heritage Trust

For all your engine's fuelling requirements, come to the specialists

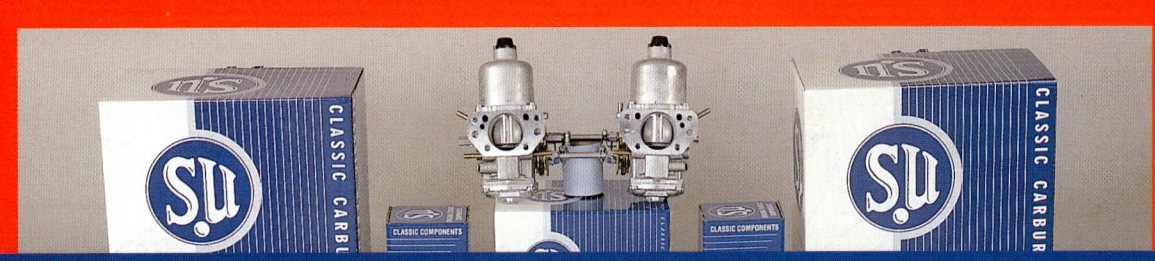

BURLEN FUEL SYSTEMS

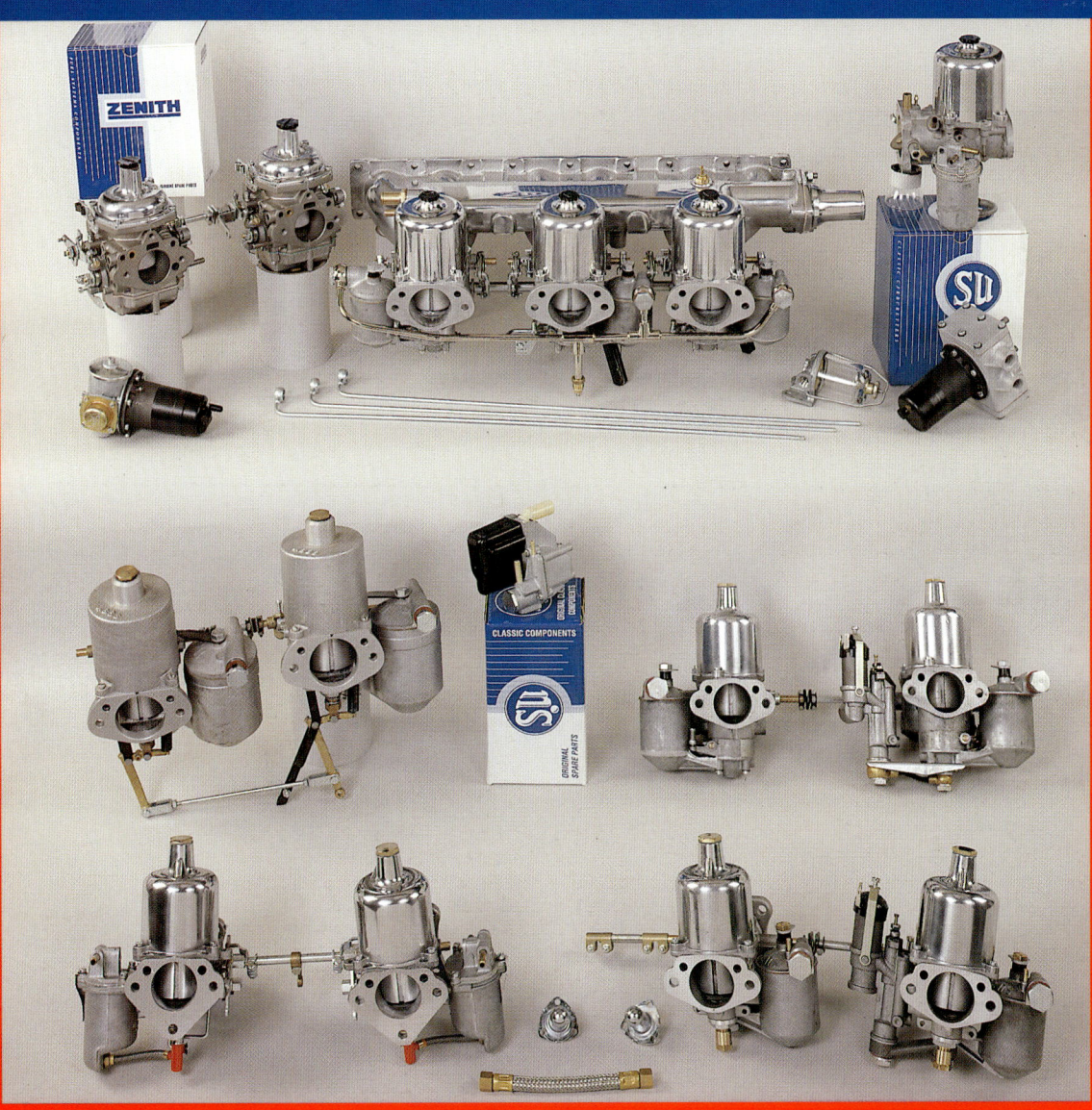

All items on this page are **GENUINE SU** and **ZENITH** carbs and spares in current production.
Virtually every part for every SU or Zenith carburettor fitted to the XK engine is available.

- 1 HIF Carbs (XJ6) ● 2 Stromberg Carbs (American Specification) ● 3 Jaguar 'E' type 4.2 Carbs, Linkage and Manifold ● 4 HS8 Carb (XJ6) ● 5 AZX 1308 Fuel Pump
- 6 Overflow Pipes ● 7 Fuel Filter ● 8 LCS Type Fuel Pump ● 9 H8 Carbs ('C' Type) ● 10 AED Automatic Choke ● 11 HD6 Carbs (3.4, 3.8 etc.) ● 12 HS6 (240)
- 13 H6 Thermo (XK 120) ● 14 Thermo Choke and Radiator Fan Switches (Otter Type) ● Braided Fuel Hose

BFS

For complete carburetters and pumps, rebuild kits, service kits and spare parts, contact us now.

☎ **(01722) 412500 Fax (01722) 334221**

Spitfire House, Castle Road, Salisbury, Wiltshire England SP1 3SA

Worldwide Mail Order ● Access/Barclaycard accepted ● Trade Enquiries Welcome

Overhauling SU carbs

Jim Patten helps your Jaguar breathe better

There's one common thread running right through the entire SS/Jaguar production span - SU carburettors. True, other makes have been used within the range (R.A.G., Solex, Weber) but from the early days right up to the demise of the Limousine, the instruments originating with Skinner's Union has predominated.

Of course today, Jaguars are intelligent beings, their engines possessed of a brain. The fuel is metered to injection points on demand with metering controlled to suit every eventuality - even the gearchanging of an automatic gearbox. So how then did the cars of the past manage? Were they just brainless twits muddling through or was their simplicity just as successful?

How the SU works

The SU carburettor is of the variable choke type and although basic in principle, is indeed extremely efficient in practice. When the throttle pedal is operated, the butterfly on the engine side of the carburettor is opened. This allows the depression in the inlet manifold (the sucking of the pistons) to draw air through the carburettor choke. The suction is sufficient to lift the carburettor piston which in turn takes the needle with it, rising from the jet below. Fuel enters the airflow and atomises, ready to quench the thirst in the waiting combustion chamber. The more the butterfly is opened, the more airflow and *vice versa*.

Above the piston is a spring and a damper. The spring acting with the weight of the piston determines its lift while the damper/plunger, which is oil-filled, restricts the lift giving satisfactory 'pick up' but will fall freely when the throttle is closed.

Seemingly, whatever state the carburettors are in, the car will still run. But just how well? That ominous smell of petrol, a reluctance to start, poor idling...

1 Remove the throttle intermediate lever from between the carburettors.

2 Slacken both couplings on the throttle linkage and remove. The carburettors will now separate.

3 On the right hand carburettor only, remove the two banjo bolts securing the starter carburettor and place this aside to be attended later.

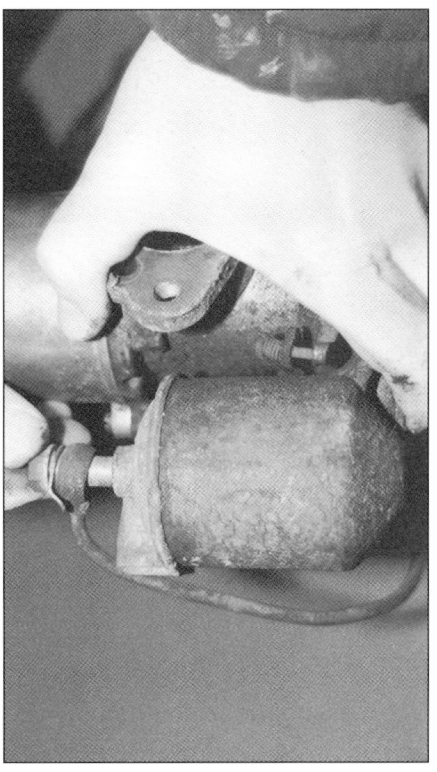

4 Remove the bolt securing the float chamber lid and lift away with the overflow pipe. Lift the lid and remove the float from inside.

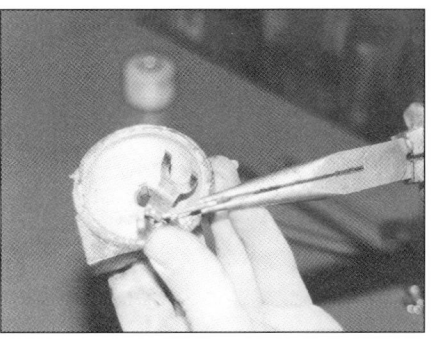

5 Pull out the knurled pin supporting the fuel level lever and remove both and the float needle valve.

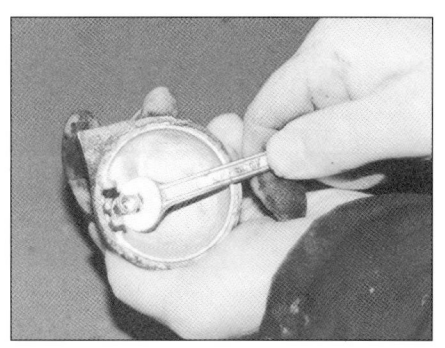

6 Now the brass valve housing can be removed.

7 With the three screws holding the dashpot undone and the dashpot removed, the suction chamber piston and spring can be lifted out.

8 Remove the four screws securing the float chamber to the main body...

9 ...this will reveal the jet/daiphragm and lower spring. It was a surprise to find the spring complete as these usually corrode and collapse.

10 Remove the jet bearing by undoing the outer locking nut.

11 Before the throttle butterfly can be removed, the two split screws should be squeezed together...

12 ... and the screws undone. The butterfly will then slide from the slot in the throttle spindle.

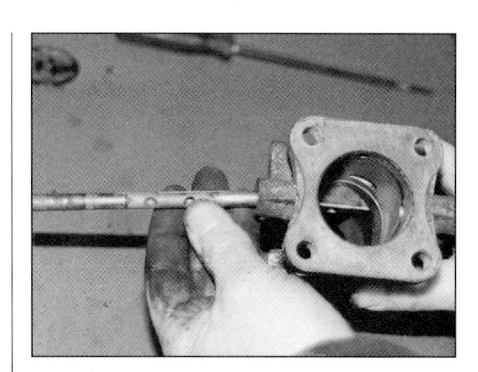

13 The spindle will now simply pull out.

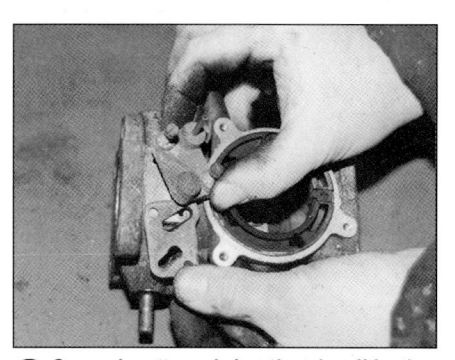

14 One carburettor only has the take off for the advance/retard pipe. If it looks a little extravagant for the job, it's that SU build the bodies for many applications.

15 Unscrew the slow running valve with its shake-proof spring.

16 The choke solenoid is retained by a clip secured with a screw. With this undone, the solenoid with its disc plunger and spring can be placed aside.

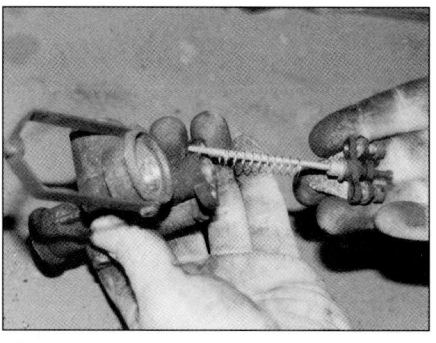

17 The main needle and spring is accessed by undoing the two retaining screws. Note the dust sheild under the screws.

18 Use a deep thin socket to remove the jet.

We stripped a pair of HD6s (one of the most common on Jaguars) and took the bundle of bits to Burlen Fuel Supplies to find out just what is involved. Good news for the home builder - they can supply every little part needed for any carburettor in the range.

The set homework until the next issue is to take the pile of bits and pieces, and clean and polish them until they are like new. (Please don't polish the inside of the pistons.) Don't worry so much about the spindles because unless they show no signs of wear, they will be replaced. Just time for a bit of idle gossip. Jaguar's choice of spelling is CARBURETTER, while the rest of us stay with CARBURETTOR. Chambers state that either will do so please, stay that letter of chastisement.

Our thanks go to Burlen Fuel Systems Ltd, Spitfire House, Castle Road, Salisbury, Wiltshire SP1 3SA. (Tel: 0722 412500; Fax: 0722 334221.)

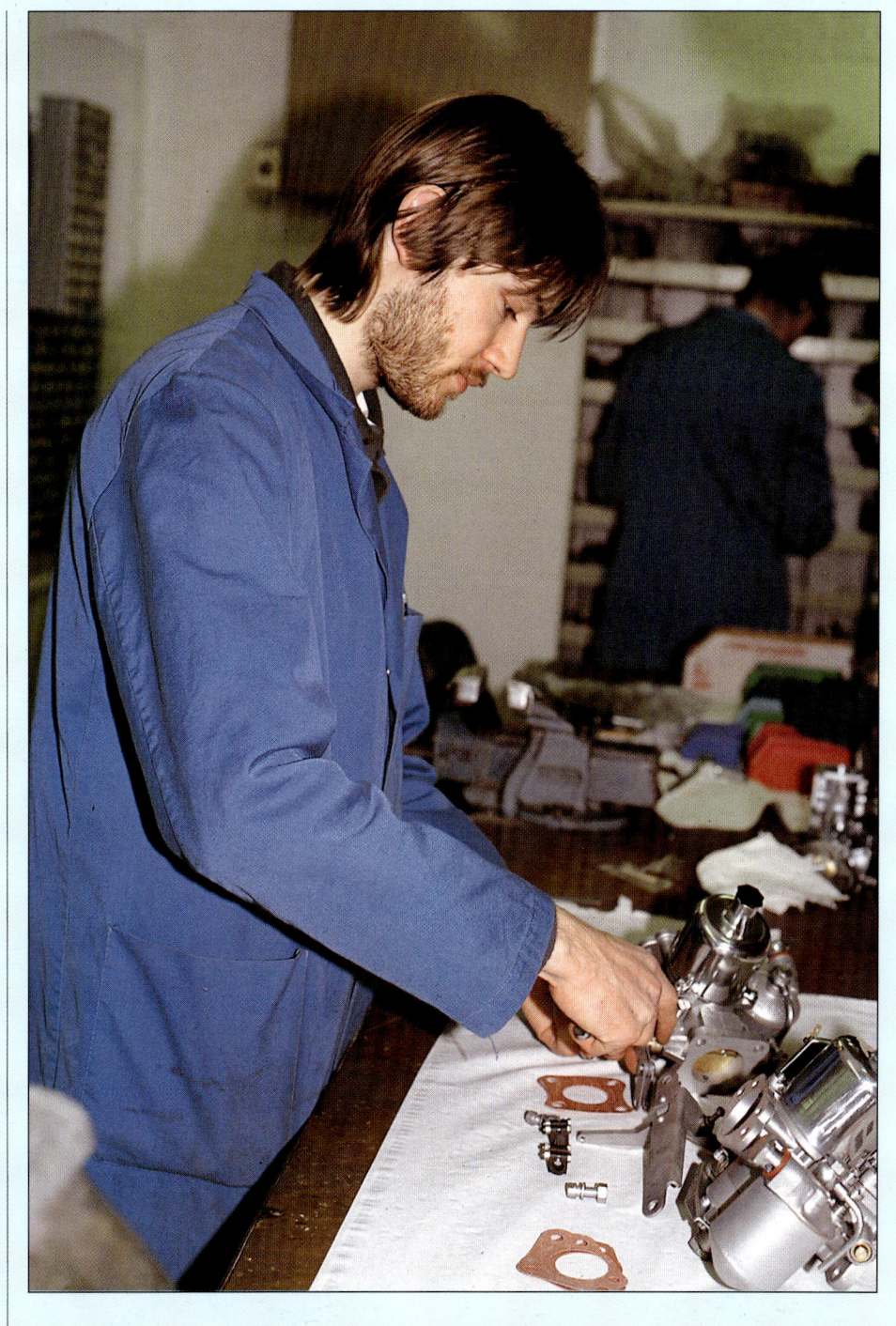

Overhauling SU carbs

PART 2

Jim Patten relates how a pair of HD6s get their breath back

Oh, to bathe in the luxury of modern equipment and purpose-designed tools. If I were cleaning these SU carburettors at home the bits would be strewn across my workbench and I would be spending hours with a toothbrush cleaning away all the grime.

There's nothing wrong in that; the end result would be the same. It was just a pleasure to be spared the tedium while Rob Harris, the restoration shop manager at Burlen Fuel Services, popped the encrusted aluminium castings into their cleaning soup and had them polished afterwards. No favours for the press here: the line-up of gleaming 'customer' castings confirmed that all parts receive the same treatment.

Rob was to show us just how the carburettors should go back together on his clinical bench. I guess that's the first lesson he could teach us before he even picked up a spanner. His bench is spotlessly clean and large enough to contain every part that he is working on with space for tools and spares. Easily duplicated at home is his use of a large spread of cardboard over the workbench to provide a clean and pleasant base to work from, allowing you to see all the parts clearly and generally making the job more straightforward.

At Burlen they have the full manufacturer's specification sheet for every carburettor. We have cribbed the relevant sections and relay the results for you here. So let's have a look over Rob's shoulder and see what he makes of our carbs.

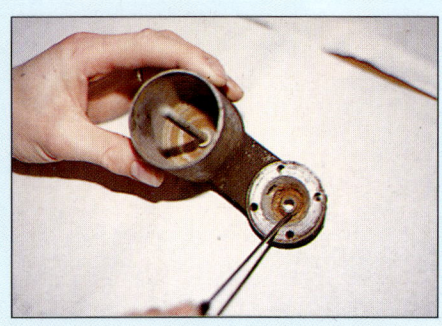

1 Not everything could be salvaged from our 'pile of bits'. One victim was this float chamber. Water and sediment often sit in the area beneath the seat of the diaphragm spring. The housing can corrode and even the spring does not escape - it just pickles and collapses. This one went in the bin in favour of another.

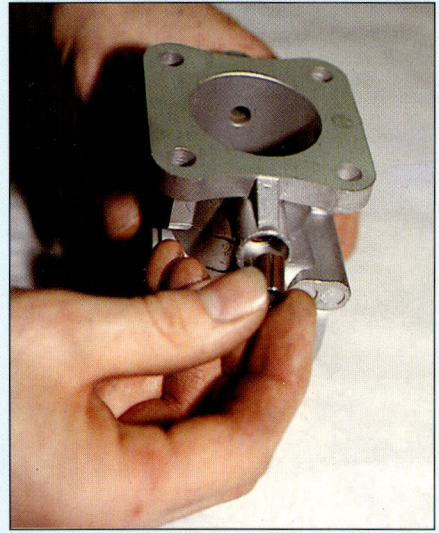

2 The spindle turns in a bush at each end of the body. They must be pushed in so that they are true to each other. First offer the bush to the body; it will nip in place due to a chamfered edge...

3 ...then, using a piece of brass scrap (Rob fittingly uses an abandoned carb. brass ferule), place the body in a vice with a soft layer on the jaws (a sheet of aluminium suitably bent is fine) and squeeze the bush in place. If you think it's going in at a funny angle, stop, check and, if necessary, start again.

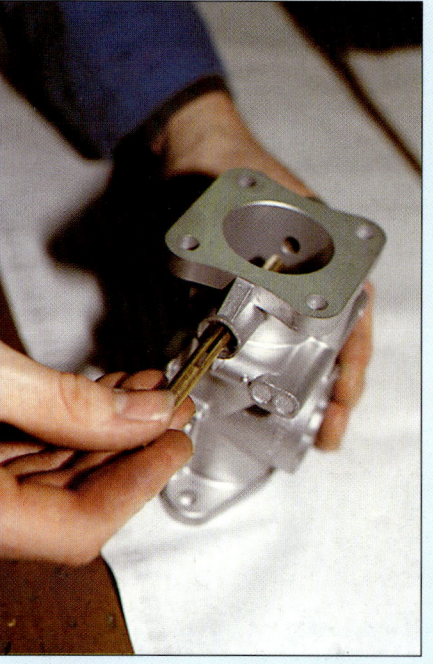

4 Slide the throttle spindle through each bush and make sure that it rotates cleanly without snatching. The long side of the spindle should be to the left-hand side of the carburettor body viewed from the back.

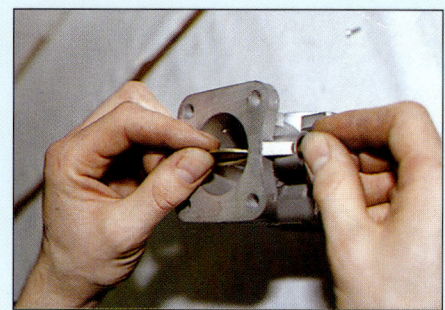

5 When the butterfly disc goes into its slot, make sure that the chamfered edge is the correct way round. It should fit snugly to the body when the throttle spindle is closed.

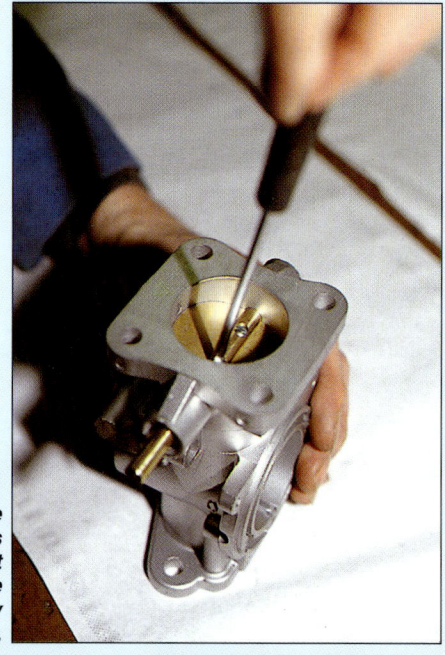

6 Two split screws keep the butterfly disc in place. Screw these fully home and then SLIGHTLY open up the split to the back. Note that the screws always face with their heads outwards. Remember, these are screws, not split pins. Always use NEW screws. If the 'split' end were to fatigue and drop off, its only exit route would be straight into the engine.

7 At each end of the spindle on the outside of the body is the spindle seal assembly. It is fitted in this order: cork gland, dished washer, gland spring and retaining endcap. Back to the vice with its protected jaws. Using another piece of soft scrap, carefully ease the seal assembly into place. Repeat for the other side.

Overhauling SU carbs

8 Refit the idle screw with a new washer and gland. Screw in until it just touches the base (do not force) and then back off 3½ turns. This will give a basic setting.

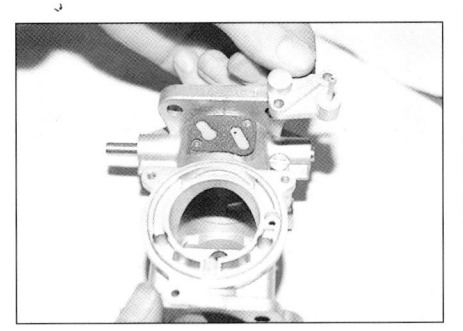

9 Using a new gasket, replace the ignition vacuum take-off plate.

10 So that the needle is in the correct position in relation to the jet, it should be fitted so that the shoulder on the base of the needle is level with the bottom of the piston. Only when this has been achieved can the grub screw on the side be tightened.

11 Smear some light oil around the piston edge and then fit the large piston spring with the colour code (in this case red) to the base. Offer the piston and spring into the dashpot (suction chamber) and check for ease of movement up and down. Fit to the carburettor body and fasten with the three retaining screws. Be extremely careful of the vulnerable needle sticking out from beneath.

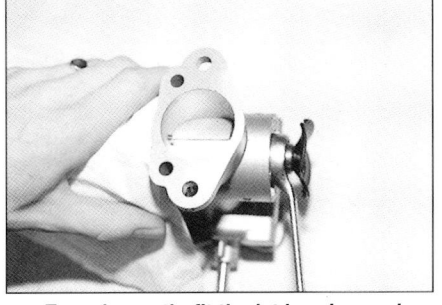

12 From beneath, fit the jet bearing and locking screw but only finger-tight at this stage. Then temporarily offer up the jet assembly. To centre the jet, lift the piston and allow it to fall making sure that the jet remains fixed. If it falls with one easy movement, fully tighten the jet bearing locking screw. Lift and let fall the piston once more to confirm that it has not moved. If the piston does not fall easily, slacken the locking screw and allow the piston to fall again. You may have to rotate the jet bearing until the best position is achieved.

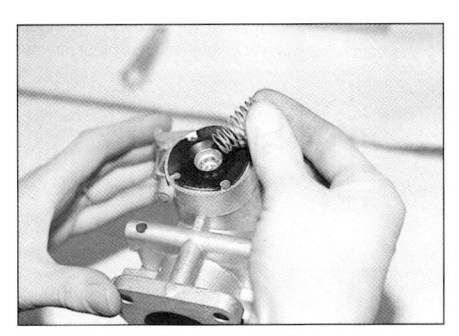

13 Withdraw the jet assembly and fit the jet housing, replacing the jet assembly after. Place on a new jet return spring. Put to one side while the float chamber is assembled.

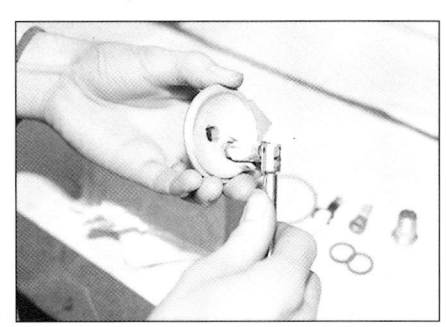

14 Using a new needle and seat kit, fit it to the float chamber lid.

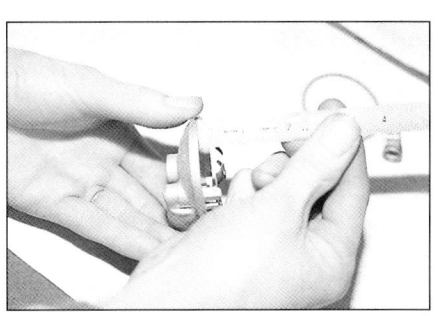

15 Fit the float level lever into the cap using a new pin. The distance between the centre of the lever and the bottom edge of the cap should be 7/16in (11.1mm). If adjustment is needed, bend the lever at the end of the straight section before it goes into the bend. An alternative way to measure the gap is to pass a bar of 7/16in diameter through the gap between the lever and the cap base.

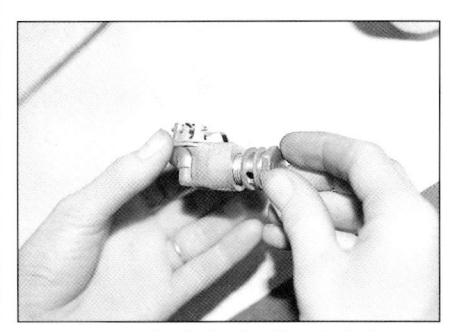

16 Temporarily fit the fuel banjo bolt, not forgetting the all-important conical filter.

17 Fit the float, cap and washer. The cap is fastened with a vent pipe between cap and bolt. Two washers are fitted each side of the bolt (the top one is plain while the bottom is vented). The vent pipe here is shortened for convenience of photography. In reality it is much longer and usually fastened to a clip on the engine block.

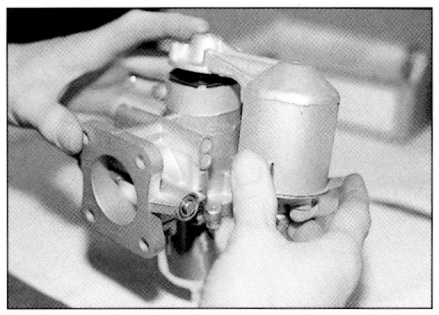

18 The float chamber can now be fitted to the main carburettor body, securing it with four screws. On cars fitted with an auxiliary starting carburettor, the support bracket should be fitted to the right-hand carburettor.

19 Fit the mixture adjustment screw with its anti-shake spring to the jet lever. Remove the dashpot, piston and spring. Look at the inside where the jet slides in the jet bearing. Turn the mixture screw until the two are level. Then, lower the jet (clockwise) 2½ turns. Refit the dashpot, piston and spring.

20 Refit the jet to the base of the auxiliary starting carburettor. Insert the needle assembly with spring and secure with the two screws, not forgetting to place the dust shield to the outside.

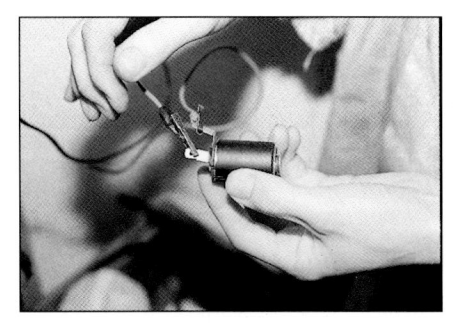

21 Check the solenoid for operation by connecting the two terminals to a 12 volt supply and observe the plunger. It should be positively raised pulled in magnetically.

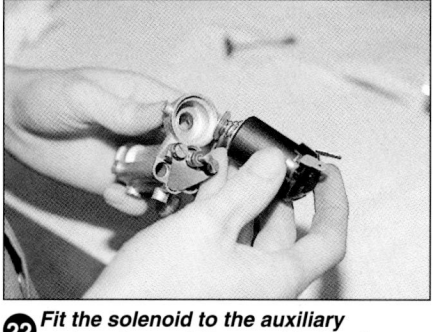

22 Fit the solenoid to the auxiliary carburettor, put the moulded end cap in place and flip the retaining clamp over the top. The clamping screw should just nip the moulded cap. Too much pressure and it will most certainly break.

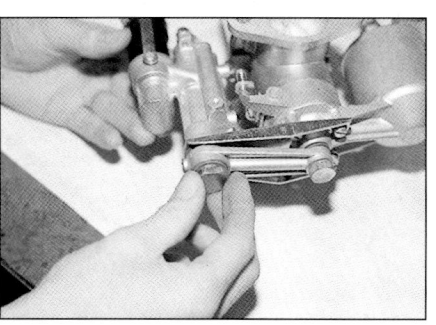

23 Connect the bridge pipe between the carburettor and auxiliary carburettor, being sure to put a seating washer on each mating surface.

There's no peculiar mystery attached to rebuilding these SUs as can be seen from the foregoing text. Every single part is available from Burlen Fuel Services and, if you don't feel up to the job, they will recondition your existing units. Repair kits can be bought that cover every aspect of the work carried out here. Try to use the correct grade of oil in the dashpots

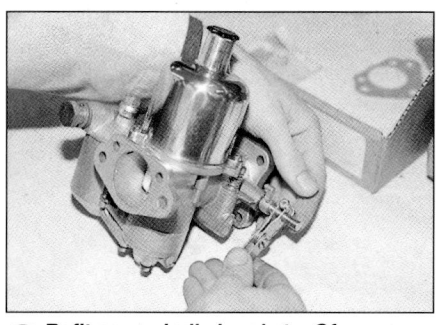

24 Refit any spindle brackets. Of course you would have made a note of these before you removed them - wouldn't you? Their final position will be determined when the carburettors are back on the car.

(available from Burlen or through outlets) or light grade machine oil if you have to.

Observe the way everything goes back. If not, it will be to your cost. An 'E' type I owned some years ago received some attention around the float chamber. During reassembly two plain washers were used to secure the cap, omitting the vented washer. Of course, the car came to a stop as the fuel would not pressurise. Unaware of the problem, I went through the ignition circuit first before eventually finding the fault a long way down the check list.

I am indebted to Rob Harris at Burlen Fuel Services for all his help with this feature. Catalogues and price lists are available from BFS Ltd, Spitfire House, Castle Road, Salisbury, Wiltshire SP1 3SA (Tel: 0722 412500; Fax: 0722 334221).

In the next issue we start a new series on rebuilding the Mk 2 series front suspension assembly.

SU APPLICATIONS, JAGUAR

Model	Year	Type
1.5-litre	1937-38	D4
1.5-litre	1939-40 1946-47	H4
2.5-litre	1936-7	HV3 or HV3 Thermo
2.5-litre competition	1936	HV4
2.5-litre	1937-40 1946-49	H3 Thermo
3.5-litre	1937-40 1946-50	H4 Thermo
Mk VII 3.4	1951-56	H6 Thermo
XK 120 3.4	1949-54	H6 Thermo
XK 120 'C'	1952	H8
XK 140 3.4	1954-57	H6 Thermo
XK 140 SE 'C' (carbs optional) (above offered on Mk VII)	1954-57	H8
2.4-litre (3rd tune option)	1956	HD6 Thermo
Mk 1 3.4-litre Mk2 3.4 & 3.8, 340 Mk VIII & IX, XK150 3.4 & 3.8 'S' type	All	HD6 Thermo

Model	Year	Type
XK 150S, 3.8 & 4.2 Mk X, 420, 420G. XJ6 (until Mar '91)	All	HD8 Thermo
'E' type 3.8 & 4.2 (excluding US emission cars)	All	HD8
240	All	HS6
XJ6 4.2 (including later 2.8) up to Series 3	All	HS8 AED
XJ6 3.4	All	HS6 AED
'E' type S3 & XJ12 (carb model) SU conversion	All	HIF6

Note that the carburettor type relates to the 'family'. There are further differences within that range depending on the application. The SU conversion for the V12 engine was never a factory option.

Footman James & Company Limited
INSURANCE HOTLINE
TEL: 021 561 4196
FAX: 021 559 9203

JAGUAR ENTHUSIASTS' CLUB

The Jaguar Enthusiasts' Club offers so much more to all Jaguar enthusiasts:

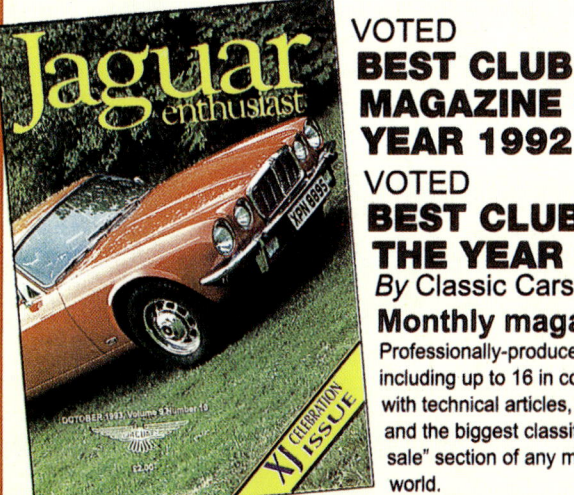

VOTED BEST CLUB MAGAZINE OF THE YEAR 1992
VOTED BEST CLUB OF THE YEAR 1993
By Classic Cars magazine

Monthly magazine
Professionally-produced, A4 80 pages, including up to 16 in colour, packed with technical articles, help and advice and the biggest classified "Jaguars for sale" section of any magazine in the world.

E-TYPE REFURBISHMENT
The total renovation by the Club of Jaguar's own Heritage V12 Series 3 E-Type – the ex-'press car' that was featured in many magazine road tests. Work to be covered in serialised detail in the Club magazine *Jaguar Enthusiast* will include bodywork, paintwork and all mechanics. Due to start shortly.

XK 120 RESTORATION
Detailed step-by-step coverage currently being serialised in *Jaguar Enthusiast*.

XJS RESTORATION
Total restoration currently being serialised in *Jaguar Enthusiast*.

Other excellent club services include:

The Club's magazine alone is well worth the subscription but our other excellent services include:

Jaguar Technical Advice
Available Mon-Sat, 10am-8pm on a mobile 'phone.

Spares Remanufacture
As well as many special Jaguar tools the Club remanufactures many vital spare parts.

Jaguar Insurance
Club scheme with special discounts only available to Club members, caters for all Jaguar/Daimler models **over five years old**.

Books & Regalia
Virtually every Jaguar book ever published available, plus a very wide range of regalia and accessories to choose from.

Specialists & Services
A booklet containing over 200 specialists listed by services they offer, county and alphabetically.

National & Local Jaguar Shows
Wide range to choose from, plus local meetings organised by over 60 Regions nationwide.

SPECIAL JAGUAR TOOLS
The following is a selection of just some of the special tools remanufactured by the Club:

Camshaft Setting Plate 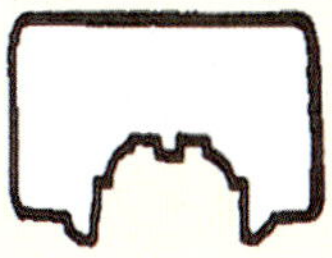
The type as supplied with the original Jaguar tool kits.
£10.75 inc VAT + 75p p&p.

Jaguar Engine Lifting Bracket 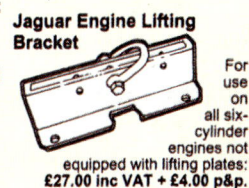
For use on all six-cylinder engines not equipped with lifting plates:
£27.00 inc VAT + £4.00 p&p.

Jaguar Timing Chain Adjuster
Made from the best quality steel, with hardened pins:
£7.20 inc VAT + £1 p&p

Jaguar Clutch Alignment Tool
A vital tool for clutch fitting: 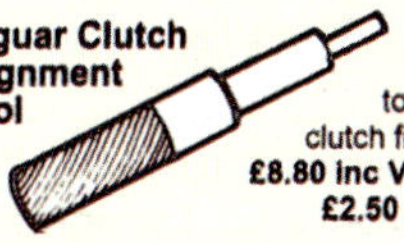
£8.80 inc VAT + £2.50 p&p.

MK1 JAGUAR RESTORATION
Coming soon: A step-by-step serialised guide in the Club's *Jaguar Enthusiast* magazine. Due to start Spring 1995.

Post to: **Jaguar Enthusiasts' Club Ltd, FREEPOST (BS-6705), Patchway, Bristol BS12 6BR**

I wish to join the Jaguar Enthusiasts' Club and I enclose my cheque for £25 (£20, plus £5 joining fee), £30 overseas – annual subscription £_____ *(payable to Jaguar Enthusiasts' Club Ltd)*

(Or) I wish to pay by Access ☐ / Visa ☐ Mastercard ☐

No_____

Expiry date/Signature_____

Name_____
PLEASE PRINT CAREFULLY

Address_____

_____ Post Code _____

Model of Jaguar_____ JW10/94

Other Books avaiable in this Series from Kelsey Publishing

E-TYPE JAGUAR RESTORATION

Step-by-step DIY restoration guide. Over 100 pages (most in colour).
£12.95 + £1 p&p (UK)

Mk2 JAGUAR RESTORATION

Step-by-step DIY restoration guide. Over 100 pages (most in colour).
£12.95 + £1 p&p (UK)

JAGUAR XJ6 RESTORATION

COMING SOON

The first fully documented full restoration on this famous Jaguar model. 124 pages (many in colour). **£14.95 + £1 p&p (UK)**

PANEL BEATING AND PAINT REFINISHING

Best illustrayed guide to panel beating plus paint refinishing. Over 100 pages.
£10.95 + £1 p&p (UK)

WELDING TECHNIQUES & WELDERS

A book which includes all the welding methods available to the DIY enthusiast and how to use them, with emphasis on MIG welding. 70 pages. **£7.95 + £1 p&p (UK)**

CLASSIC TRIM

Step-by-step DIY restoration guide for classic car trim. Over 70 pages, many in colour.
£10.95 + £1 p&p (UK)

Post coupon to: Books Department, Kelsey Publishing Ltd., Kelsey House, 77 High Street, Beckenham, Kent BR3 1AN. Tel: 0181 658 3531

- E Type Jaguar **£13.95 inc P&P** ☐
- Mk 2 Jaguar Restoration **£13.95 inc P&P** ☐
- Jaguar XJ6 Restoration **£15.95 inc P&P** ☐
- Panel Beating & Paint Refinishing **£11.95 inc P&P** ☐
- Welding Techniques & Welders **£8.95 inc P&P** ☐
- Classic Trim **£11.95 inc P&P** ☐

I enclose my cheque for £............... *(payable to Kelsey Publishing)*

(Or) I wish to pay by Access ☐ /Visa ☐

Number ....

Expiry Date .... Signature ....

Overseas ADD £2.00 per book for surface mail (£4.00 Airmail)

Name ....

Address ....
....
....

Post Code ....   JW 6

VSE

JAGUAR

20 YEARS SERVICE TO THE TRADE

REALISTIC PRICES TO THE ENTHUSIAST

Specialist engine reconditioners and parts suppliers

WARNING

DO NOT REBUILD YOUR JAGUAR ENGINE...

...without first obtaining your FREE engine parts and information book from **VSE** *(trade and export welcome)*

VSE stock a full range of engine components for Jaguar 6 cylinder engines at substantial discounts including; pistons, rings, bearings, gaskets, seals, camshafts, conrods, bushes, ring gears, valves, guides, cam followers etc, etc., plus all the important lock tags, bolts etc.

VSE also offer a full range of machining services including; rebores, crack testing, crack repairs and crack grinding, balancing, valve guide and seat fitting, surfacing etc, etc, all at realistic prices.

VSE also stock a range of exchange, cylinder heads, engines, cranks, rods, etc.

VSE can naturally rebuild your own engine, tuning, conversions, lead free, dyno testing. In fact, for anything to do with engines contact **VSE**.

VSE
Llanbister, Llandrindod Wells, Powys
LD1 6TL
Tel: 01597 840 308 / 0831 280157
Fax: 01597 840 661